THE DRAGONS
OF EDEN

THE DRAGONS OF EDEN

*Speculations on
the Evolution of Human
Intelligence*

CARL SAGAN

For my wife, Linda, with love

Mankind is poised midway between the gods and the beasts.

PLOTINUS

The main conclusion arrived at in this work, namely, that man is descended from some lowly-organized form, will, I regret to think, be highly distasteful to many persons. But there can hardly be a doubt that we are descended from barbarians. The astonishment which I felt on first seeing a party of Fuegians on a wild and broken shore will never be forgotten by me, for the reflection at once rushed into my mind— such were our ancestors. These men were absolutely naked and bedaubed with paint, their long hair was tangled, their mouths frothed in excite- ment, and their expression was wild, startled, and distrustful. They possessed hardly any arts, and, like wild animals, lived on what they could catch; they had no government, and were merciless to everyone not of their own small tribe. He who has seen a savage in his native land will not feel much shame, if forced to acknowledge that the blood of some more humble creature flows in his veins. For my own part, I would as soon be descended from that heroic little monkey, who braved his dreaded enemy in order to save the life of his keeper; or from that old baboon who, descending from the mountains, carried away in tri- umph his young comrade from a crowd of astonished dogs—as from a savage who delights to torture his enemies, offers up bloody sacrifices, practices infanticide without remorse, treats his wives like slaves, knows no decency, and is haunted by the grossest superstitions.

Man may be excused for feeling some pride at having risen, though not through his own exertions, to the very summit of the organic scale; and the fact of his having thus risen, instead of having been aborigi- nally placed there, may give him hopes for a still higher destiny in the distant future. But we are not here concerned with hopes or fears, only with the truth as far as our reason allows us to discover it. I have given the evidence to the best of my ability; and we must acknowledge, as it seems to me, that man with all his noble qualities, with sympathy which feels for the most debased, with benevolence which extends not only to other men but to the humblest living creature, with his godlike intellect which has penetrated into the movements and constitution of the solar system—with all these exalted powers—Man still bears in his bodily frame the indelible stamp of his lowly origin.

CHARLES DARWIN
The Descent of Man

I am a brother to dragons, and a companion to owls.
Job 30:29

CONTENTS

FOREWORD

Carl Sagan's *The Dragons of Eden* was published in the spring of 1977, and by mid-June it had earned a spot on the *New York Times* bestseller list, debuting at number five (just below Alex Haley's *Roots*, but two slots above *The Book of Lists*, which also made its first appearance that same week). Sagan's wide-ranging "speculations on the evolution of human intelligence" clearly struck a chord with a broad audience, for his scientifically erudite, elegantly composed essay in intellectual curiosity remained on the *Times* chart for more than thirty weeks. Common readers were not alone in their admiration for *The Dragons of Eden*, and the book's popular appeal was complemented by the award of the 1978 Pulitzer Prize for General Nonfiction.

Sagan (1934-1996) was a scientist of renown and some celebrity at the time *The Dragons of Eden* was published. Professor of Astronomy and Space Sciences at Cornell University, where he was also Director of the Laboratory for Planetary Studies, he had earned a reputation as an iconoclast by challenging, early in his career (and correctly, as it turned out), standard scientific explanations of the changing surface light of Mars, the Venus greenhouse effect, and other astronomical matters. His interest in extraterrestrial intelligence had led to several books, while his infectious sense of wonder and his gift for graceful, inspired explication had already made him a noted popularizer of scientific and cosmological concepts—leading John Updike, in his review of *The Dragons of Eden*, to sniff that Sagan "enjoys appearing on *The Tonight Show*, exposing science to the fans of Johnny Carson." Other critics were more generous with regard to the professor's television appearances. Describing a long monologue—a "cosmological crash course"—on the evolution of the Earth that Sagan launched into for the benefit of Carson's insomniacs in the early 1970s, Stuart Bauer wrote that "one was willing to bet that if a million teenagers had been watching, at least a hundred thousand vowed on the spot to become an astronomer like him." (Indeed, three years after *The Dragons of Eden* was published, Sagan's eight-million-dollar television spectacular *Cosmos* reached an international audience of four hundred million viewers, becoming the most highly rated series in public television history, unsurpassed until the airing of Ken Burns's *The Civil War* a decade later.)

While he had previously contributed the entry on "Life" to the *Encyclopedia Britannica*, *The Dragons of Eden* was Sagan's first book to venture freely beyond the realm of astronomy. Yet his intentness on seeing the evolution of human intelligence in a cosmological perspective gives his ruminations on his theme a refreshing breadth of import and attention. His frequent, often exhilarating segues from anatomical and biological detail to unexamined elements of enduring myths—the expulsion from Eden, or St. George's battle with a dragon—entwine human and natural history in provocative and illuminating ways. As Stephen Toulmin wrote in *The New York Review of Books* when *The Dragons of Eden* first appeared, the volume reveals Sagan to be a "true 'natural philosopher,' whose concern with extraterrestrial intelligence is only one element in a larger scientific program, and whose real goal is to produce a revised version of the story of human history and destiny, within the boundary conditions set by the ideas of twentieth-century natural science." As such, Sagan's enterprise continues a cosmological tradition that reaches back beyond classical antiquity.

Equally at ease with the scientific method and the literary imagination, with experiment and metaphor (myth being, in his deft description, "a metaphor of some subtlety on a subject difficult to describe in any other way"), Sagan is able to sketch a vision bold enough to engage in all its richness and perplexities the large question of what it means to be human, and to spice his speculations with a playfulness of idea and expression that increases the reader's appetite for intellectual adventure. (His concentration of the history of the universe into the time span of a single year is worth the price of the book all by itself.)

The Dragons of Eden was written more than a quarter-century ago, and surely some of its theorizing has been overtaken by new research and discovery. And yet the vivid sweep of the book's passionate perception still carries the reader along today, opening unforgettable vistas of knowledge and imagination. What Stephen Toulmin concluded in The New York Review in 1977 still applies: "One may find it necessary to take issue with Carl Sagan over many details, but the overall force of his argument will survive a great deal of correction of the details. It is a pleasure to have someone writing about scientific cosmology once again so elegantly, intelligently, and with such literary flair."

James Mustich, Jr.
June 2004

*In good speaking, should not the
mind of the speaker know the truth of the
matter about which he is to speak?*

PLATO
Phaedrus

*I do not know where to find in any literature,
whether ancient or modern, any adequate account
of that nature with which I am acquainted.
Mythology comes nearest to it of any.*

HENRY DAVID THOREAU
The Journal

INTRODUCTION

JACOB BRONOWSKI was one of a small group of men and women in any age who find all of human knowledge—the arts and sciences, philosophy and psychology—interesting and accessible. He was not confined to a single discipline, but ranged over the entire panorama of human learning. His book and television series, *The Ascent of Man*, are a superb teaching tool and a remarkable memorial; they are, in a way, an account of how human beings and human brains grew up together.

His last chapter/episode, called "The Long Childhood," describes the extended period of time—longer relative to our lifespan than for any other species—in which young humans are dependent on adults and exhibit immense plasticity—that is, the ability to learn from their environment and their culture. Most organisms on Earth depend on their genetic information, which is "prewired" into their nervous systems, to a much greater extent than they do on their extragenetic information, which is acquired during their lifetimes. For human beings, and indeed for all mammals, it is the other way around. While our behavior is still significantly controlled by our genetic inheritance, we have, through our brains, a much richer opportunity to blaze new behavioral and cultural pathways on short time scales. We have made a kind of bargain with nature: our children will be difficult to raise, but their capacity for new learning will greatly enhance the chances of survival of the human species. In addition, human beings have, in the most recent few tenths of a percent of our existence, invented not only extragenetic but also extrasomatic knowledge: information stored outside our bodies, of which writing is the most notable example.

The time scale for evolutionary or genetic change is very long. A characteristic period for the emergence of one advanced species from another is perhaps a hundred thousand years; and very often the differences in behavior between closely related species—say, lions and tigers—do not seem very great. An example of recent evolution of organ systems in humans is our toes. The big toe plays an important function in balance while walking; the other toes have much less obvious utility. They are clearly evolved from fingerlike appendages for grasping and swinging, like those of arboreal apes and monkeys. This evolution constitutes a respecialization—the adaptation of an organ system originally evolved for one function to another and quite different function—which required about ten million years to emerge. (The feet of the mountain gorilla have undergone a similar although quite independent evolution.)

But today we do not *have* ten million years to wait for the next advance. We live in a time when our world is changing at an unprecedented rate. While the changes are largely of our own making, they cannot be ignored. We must adjust and adapt and control, or we perish.

Only an extragenetic learning system can possibly cope with the swiftly changing circumstances that our species faces. Thus the recent rapid evolution of human intelligence is not only the cause of but also the only conceivable solution to the many serious problems that beset us. A better understanding of the nature and evolution of human intelligence just possibly might help us to deal intelligently with our unknown and perilous future.

I am interested in the evolution of intelligence for another reason as well. We now have at our command, for the first time in human history, a powerful tool—the large radio telescope—which is capable of communication over immense interstellar distances. We are just beginning to employ it in a halting and tentative manner, but with a perceptibly increasing pace, to determine whether other civilizations on unimaginably distant and exotic worlds may be sending radio messages to us. Both the existence of those other civilizations and the nature of the messages they may be sending depend on the universality of the process of evolution of intelligence that has occurred on Earth. Conceivably, some hints or insights helpful in the quest for extraterrestrial intelligence might be derived from an investigation of the evolution of terrestrial intelligence.

I was pleased and honored to deliver the first Jacob Bronowski Memorial Lecture in Natural Philosophy in November 1975, at the University of Toronto. In writing this book, I have expanded substantially the scope of that lecture, and have been in return provided with an exhilarating opportunity to learn something about subjects in which I am not expert. I found irresistible the temptation to synthesize some of what I learned into a coherent picture, and to tender some hypotheses on the nature and evolution of human intelligence that may be novel, or that at least have not been widely discussed.

The subject is a difficult one. While I have formal training in biology, and have worked for many years on the origin and early evolution of life, I have little formal education in, for example, the anatomy and physiology of the brain. Accordingly, I proffer the following ideas with a substantial degree of trepidation; I know very well that many of them are speculative and can be proved or disproved only on the anvil of experiment. At the very least, this inquiry has provided me with an opportunity to look into an entrancing subject; perhaps my remarks will stimulate others to look more deeply.

The great principle of biology—the one that, as far as we know, distinguishes the biological from the physical sciences— is evolution by natural selection, the brilliant discovery of Charles Darwin and Alfred Russel Wallace in the middle of the nineteenth century.* It is through natural selection, the preferential survival and replication of organisms that are by accident better adapted to their environments, that the elegance and beauty of contemporary life forms have emerged. The development of an organ system as complex as the brain must be inextricably tied to the earlier history of life, its fits and starts and dead ends, the tortuous adaptation of organisms to conditions that change once again, leaving the life form that once was supremely adapted again in danger of extinction. Evolution is adventitious and not foresighted. Only through the deaths of an immense number of slightly maladapted organisms are we, brains and all, here today.

Biology is more like history than it is like physics; the accidents and errors and lucky happenstances of the past powerfully prefigure the present. In approaching as difficult a biological problem as the nature and evolution of human intelligence, it seems to me at least prudent to give substantial weight to arguments derived from the evolution of the brain.

---

*Since the time of the famous Victorian debate between Bishop Wilberforce and T. H. Huxley, there has been a steady and notably unproductive barrage fired against the Darwin/Wallace ideas, often by those with doctrinal axes to grind. Evolution is a fact amply demonstrated by the fossil record and by contemporary molecular biology. Natural selection is a successful theory devised to explain the fact of evolution. For a very polite response to recent criticisms of natural selection, including the quaint view that it is a tautology ("Those who survive survive"), see the article by Gould (1976) listed in the references at the back of this book. Darwin was, of course, a man of his times and occasionally given—as in his remarks on the inhabitants of Tierra del Fuego quoted above—to self-congratulatory comparisons of Europeans with other peoples. In fact, human society in pretechnological times was much more like that of the compassionate, communal and cultured Bushman hunter-gatherers of the Kalahari Desert than the Fuegians Darwin, with some justification, derided. But the Darwinian insights—on the existence of evolution, on natural selection as its prime cause, and on the relevance of these concepts to the nature of human beings—are landmarks in the history of human inquiry, the more so because of the dogged resistance which such ideas evoked in Victorian England, as, to a lesser extent, they still do today.

My fundamental premise about the brain is that its workings — what we sometimes call "mind"—are a consequence of its anatomy and physiology, and nothing more. "Mind" may be a consequence of the action of the components of the brain severally or collectively. Some processes may be a function of the brain as a whole. A few students of the subject seem to have concluded that, because they have been unable to isolate and localize all higher brain functions, no future generation of neuroanatomists will be able to achieve this objective. But absence of evidence is not evidence of absence. The entire recent history of biology shows that we are, to a remarkable degree, the results of the interactions of an extremely complex array of molecules; and the aspect of biology that was once considered its holy of holies, the nature of the genetic material, has now been fundamentally understood in terms of the chemistry of its constituent nucleic acids, DNA and RNA, and their operational agents, the proteins. There are many instances in science, and particularly in biology, where those closest to the intricacies of the subject have a more highly developed (and ultimately erroneous) sense of its intractability than those at some remove. On the other hand, those at too great a distance may, I am well aware, mistake ignorance for perspective. At any rate, both because of the clear trend in the recent history of biology and because there is not a shred of evidence to support it, I will not in these pages entertain any hypotheses on what used to be called the mind-body dualism, the idea that inhabiting the matter of the body is something made of quite different stuff, called mind.

Part of the enjoyment and indeed delight of this subject is its contact with all areas of human endeavor, particularly with the possible interaction between insights obtained from brain physiology and insights obtained from human introspection. There is, fortunately, a long history of the latter, and in former times the richest, most intricate and most profound of these were called myths. "Myths," declared Salustius in the fourth century, "are things which never happened but always are." In the Platonic dialogues and *The*

Republic, every time Socrates cranks up a myth—the parable of the cave, to take the most celebrated example—we know that we have arrived at something central.

I am not here employing the word "myth" in its present popular meaning of something widely believed and contrary to fact, but rather in its earlier sense, as a metaphor of some subtlety on a subject difficult to describe in any other way. Accordingly, I have interspersed in the discussion on the following pages occasional excursions into myths, ancient and modern. The title of the book itself comes from the unexpected aptness of several different myths, traditional and contemporary.

While I hope that some of my conclusions may be of interest to those whose profession is the study of human intelligence, I have written this book for the interested layman. Chapter 2 presents arguments of somewhat greater difficulty than the rest of this inquiry, but still, I hope, accessible with only a little effort. Thereafter, the book should be smooth sailing. Occasional technical terms are usually defined when first introduced, and are collected in the glossary. The figures and the glossary are additional tools to aid those with no formal background in science, although understanding my arguments and agreeing with them are not, I suspect, the same thing.

In 1754, Jean Jacques Rousseau, in the opening paragraph of his *Dissertation on the Origin and Foundation of the Inequity of Mankind*, wrote:

> Important as it may be, in order to judge rightly of the natural state of man, to consider him from his origin … I shall not follow his organization through its successive developments. … On this subject I could form none but vague and almost imaginary conjectures. Comparative anatomy has as yet made too little progress, and the observations of naturalists are too uncertain to afford an adequate basis for any solid reasoning.

Rousseau's cautions of more than two centuries ago are valid still. But there has been remarkable progress in investigating both comparative brain anatomy and animal and human behavior, which he correctly recognized as critical to the problem. It may not be premature today to attempt a preliminary synthesis.

What seest thou else
In the dark backward and abysm of time?

WM. SHAKESPEARE
The Tempest

CHAPTER I

THE COSMIC CALENDAR

THE WORLD is very old, and human beings are very young. Significant events in our personal lives are measured in years or less; our lifetimes in decades; our family genealogies in centuries; and all of recorded history in millennia. But we have been preceded by an awesome vista of time, extending for prodigious periods into the past, about which we know little—both because there are no written records and because we have real difficulty in grasping the immensity of the intervals involved.

Yet we are able to date events in the remote past. Geological stratification and radioactive dating provide information on archaeological, paleontological and geological events; and astrophysical theory provides data on the ages of planetary surfaces, stars, and the Milky Way Galaxy, as well as an estimate of the time that has elapsed since that extraordinary event called the Big Bang—an explosion that involved all of the matter and energy in the present universe. The Big Bang may be the beginning of the universe, or it may be a discontinuity in which information about the earlier history of the universe was destroyed. But it is certainly the earliest event about which we have any record.

The most instructive way I know to express this cosmic chronology is to imagine the fifteen-billion-year lifetime of the universe (or at least its present incarnation since the Big Bang) compressed into the span of a single year. Then every billion years of Earth history would correspond to about twenty-four days of our cosmic year, and one second of that year to 475 real revolutions of the Earth about the sun. On pages 12 through 14 I present the cosmic chronology in three forms: a list of some representative pre-December dates; a calendar for the month of December; and a closer look at the late evening of New Year's Eve. On this scale, the events of our history books—even books that make significant efforts to deprovincialize the present—are so compressed that it is necessary to give a second-by-second recounting of the last seconds of the cosmic year. Even then, we find events listed as contemporary that we have been taught to consider as widely separated in time. In the history of life, an equally rich tapestry must have been woven in other periods—for example, between 10:02 and 10:03 on the morning of April 6th or

PRE-DECEMBER DATES

Big Bang	January 1
Origin of the Milky Way Galaxy	May 1
Origin of the solar system	September
Formation of the Earth	September 14
Origin of life on Earth	~September 25
Formation of the oldest rocks known on Earth	October 2
Date of oldest fossils (bacteria and blue-green algae)	October 9
Invention of sex (by microorganisms)	~November 1
Oldest fossil photosynthetic plants	November 12
Eukaryotes (first cells with nuclei) flourish	November 15

~ = *approximately*

COSMIC CALENDAR

DECEMBER

SUNDAY	MONDAY	TUESDAY	WEDNESDAY	THURSDAY	FRIDAY	SATURDAY
	1 Significant oxygen atmosphere begins to develop on Earth.	**2**	**3**	**4**	**5** Extensive vulcanism and channel formation on Mars.	**6**
7	**8**	**9**	**10**	**11**	**12**	**13**
14	**15**	**16** First worms.	**17** Precambrian ends. Paleozoic Era and Cambrian Period begin. Invertebrates flourish.	**18** First oceanic plankton. Trilobites flourish.	**19** Ordovician Period. First fish, first vertebrates.	**20** Silurian Period. First vascular plants. Plants begin colonization of land.
21 Devonian Period begins. First insects. Animals begin colonization of land.	**22** First amphibians. First winged insects.	**23** Carboniferous Period. First trees. First reptiles.	**24** Permian Period begins. First dinosaurs.	**25** Paleozoic Era ends. Mesozoic Era begins.	**26** Triassic Period. First mammals.	**27** Jurassic Period. First birds.
28 Cretaceous Period. First flowers. Dinosaurs become extinct.	**29** Mesozoic Era ends. Cenozoic Era and Tertiary Period begin. First cetaceans. First primates.	**30** Early evolution of frontal lobes in the brains of primates. First hominids. Giant mammals flourish.	**31** End of the Pliocene Period. Quaternary (Pleistocene and Holocene) Period. First humans.			

DECEMBER 31

Origin of *Proconsul* and *Ramapithecus*, probable ancestors of apes and men	~1:30 P.M.
First humans	~10:30 P.M.
Widespread use of stone tools	11:00 P.M.
Domestication of fire by Peking man	11:46 P.M.
Beginning of most recent glacial period	11:56 P.M.
Seafarers settle Australia	11:58 P.M.
Extensive cave painting in Europe	11:59 P.M.
Invention of agriculture	11:59:20 P.M.
Neolithic civilization; first cities	11:59:35 P.M.
First dynasties in Sumer, Ebla and Egypt; development of astronomy	11:59:50 P.M.
Invention of the alphabet; Akkadian Empire	11:59:51 P.M.
Hammurabic legal codes in Babylon; Middle Kingdom in Egypt	11:59:52 P.M.
Bronze metallurgy; Mycenaean culture; Trojan War; Olmec culture: invention of the compass	11:59:53 P.M.
Iron metallurgy; First Assyrian Empire; Kingdom of Israel; founding of Carthage by Phoenicia	11:59:54 P.M.
Asokan India; Ch'in Dynasty China; Periclean Athens; birth of Buddha	11:59:55 P.M.
Euclidean geometry; Archimedean physics; Ptolemaic astronomy; Roman Empire; birth of Christ	11:59:56 P.M.
Zero and decimals invented in Indian arithmetic; Rome falls; Moslem conquests	11:59:57 P.M.
Mayan civilization; Sung Dynasty China; Byzantine empire; Mongol invasion; Crusades	11:59:58 P.M.
Renaissance in Europe; voyages of discovery from Europe and from Ming Dynasty China; emergence of the experimental method in science	11:59:59 P.M.
Widespread development of science and technology; emergence of a global culture; acquisition of the means for self-destruction of the human species; first steps in spacecraft planetary exploration and the search for extraterrestrial intelligence	Now: The first second of New Year's Day

September 16th. But we have detailed records only for the very end of the cosmic year.

The chronology corresponds to the best evidence now available. But some of it is rather shaky. No one would be astounded if, for example, it turns out that plants colonized the land in the Ordovician rather than the Silurian Period; or that segmented worms appeared earlier in the Precambrian Period than indicated. Also, in the chronology of the last ten seconds of the cosmic year, it was obviously impossible for me to include all significant events; I hope I may be excused for not having explicitly mentioned advances in art, music and literature or the historically significant American, French, Russian and Chinese revolutions.

The construction of such tables and calendars is inevitably humbling. It is disconcerting to find that in such a cosmic year the Earth does not condense out of interstellar matter until early September; dinosaurs emerge on Christmas Eve; flowers arise on December 28th; and men and women originate at 10:30 P.M. on New Year's Eve. All of recorded history occupies the last ten seconds of December 31; and the time from the waning of the Middle Ages to the present occupies little more than one second. But because I have arranged it that way, the first cosmic year has just ended. And despite the insignificance of the instant we have so far occupied in cosmic time, it is clear that what happens on and near Earth at the beginning of the second cosmic year will depend very much on the scientific wisdom and the distinctly human sensitivity of mankind.

What the hammer? What the chain?
In what furnace was thy brain?
What the anvil? What dread grasp
Dare its deadly terrors clasp?

WM. BLAKE
"The Tyger"

Of all animals, man has the largest brain
in proportion to his size.

ARISTOTLE
The Parts of Animals

CHAPTER II

GENES
AND BRAINS

BIOLOGICAL evolution has been accompanied by increasing complexity. The most complex organisms on Earth today contain substantially more stored information, both genetic and extragenetic, than the most complex organisms of, say, two hundred million years ago—which is only 5 percent of the history of life on the planet, five days ago on the Cosmic Calendar. The simplest organisms on Earth today have just as much evolutionary history behind them as the most complex, and it may well be that the internal biochemistry of contemporary bacteria is more efficient than the internal biochemistry of the bacteria of three billion years ago. But the amount of genetic information in bacteria today is probably not vastly greater than that in their ancient bacterial ancestors. It is important to distinguish between the amount of information and the quality of that information.

The various biological forms are called taxa (singular, taxon). The largest taxonomic divisions distinguish between plants and animals, or between those organisms with poorly developed nuclei in their cells (such as bacteria and blue-green algae) and those with very clearly demarcated and elaborately architectured nuclei (such as protozoa or people). All organisms on the planet Earth, however, whether they have well-defined nuclei or not, have chromosomes, which contain the genetic material passed on from generation to generation. In all organisms the hereditary molecules are nucleic acids. With a few unimportant exceptions, the hereditary nucleic acid is always the molecule called DNA (deoxyribonucleic acid). Much finer divisions among various sorts of plants and animals, down to species, subspecies and races, can also be described as separate taxa.

A species is a group that can produce fertile offspring by crosses within but not outside itself. The mating of different breeds of dogs yields puppies which, when grown, will be reproductively competent dogs. But crosses between species—even species as similar as donkeys and horses—produce infertile offspring (in this case, mules). Donkeys and horses are therefore categorized as separate species. Viable but infertile matings of more widely separated species—for example, lions and tigers—sometimes occur, and if, rarely, the offspring are fertile, this indicates only that the definition of species is a little fuzzy. All human beings are members of the same species, *Homo sapiens*, which means, in optimistic Latin, "Man, the wise." Our probable ancestors, *Homo erectus* and *Homo habilis*—now extinct—are classified as of the same genus (*Homo*) but of different species, although no one (at least lately) has attempted the appropriate experiments to see if crosses of them with us would produce fertile offspring.

In earlier times it was widely held that offspring could be produced by crosses between extremely different organisms. The Minotaur whom Theseus slew was said to be the result of a mating between a bull and a woman. And the Roman historian Pliny suggested that the ostrich, then newly discovered, was the result of a cross between a giraffe and a gnat. (It would, I suppose, have to be a female giraffe and a male gnat.) In practice there must be many such crosses which have not been attempted because of a certain understandable lack of motivation.

The chart that appears on page 22 will be referred to repeatedly in this chapter. The solid curve on it shows the times of earliest emergence of various major taxa. Many more taxa exist, of course, than are shown by the few points in the figure. But the curve is representative of the much denser array of points that would be necessary to characterize the tens of millions of separate taxa which have emerged during the history of life on our planet. The major taxa, which have evolved most recently, are by and large the most complicated.

Some notion of the complexity of an organism can be obtained merely by considering its behavior—that is, the number of different functions it is called upon to perform in its lifetime. But complexity can also be judged by the minimum information content in the organism's genetic material. A typical human chromosome has one very long DNA molecule wound into coils, so that the space it occupies is very much smaller than it would be if it were unraveled. This DNA molecule is composed of smaller building blocks, a little like the rungs and sides of a rope ladder. These blocks are called nucleotides and come in four varieties. The language of life, our hereditary information, is determined by the sequence of the four different sorts of nucleotides. We might say that the language of heredity is written in an alphabet of only four letters.

But the book of life is very rich; a typical chromosomal DNA molecule in a human being is composed of about five billion pairs of nucleotides. The genetic instructions of all the other taxa on Earth are written in the same language, with the same code book. Indeed, this shared genetic language is one line of evidence that all the organisms on Earth are descended from a single ancestor, a single instance of the origin of life some four billion years ago.

The information content of any message is usually described in units called bits, which is short for "binary digits." The simplest arithmetical scheme uses not ten digits (as we do because of the evolutionary accident that we have ten fingers) but only two,

0 and 1. Thus any sufficiently crisp question can be answered by a single binary digit—0 or 1, yes or no. If the genetic code were written in a language of two letters rather than four letters, the number of bits in a DNA molecule would equal twice the number of nucleotide pairs. But since there are four different kinds of nucleotides, the number of bits of information in DNA is four times the number of nucleotide pairs. Thus if a single chromosome has five billion (5×10^9) nucleotides, it contains twenty billion (2×10^{10}) bits of information. [A symbol such as 10^9 merely indicates a one followed by a certain number of zeroes—in this case, nine of them.]

How much information is twenty billion bits? What would be its equivalent, if it were written down in an ordinary printed book in a modern human language? Alphabetical human languages characteristically have twenty to forty letters plus one or two dozen numerals and punctuation marks; thus sixty-four alternative characters should suffice for most such languages. Since 2^6 equals 64 ($2 \times 2 \times 2 \times 2 \times 2 \times 2$), it should take no more than six bits to specify a given character. We can think of this being done by a sort of game of "Twenty Questions," in which each answer corresponds to the investment of a single bit to a yes/no question. Suppose the character in question is the letter J. We might specify it by the following procedure:

FIRST QUESTION: Is it a letter (0) or some other character (1)?
ANSWER: A letter (0)
SECOND QUESTION: Is it in the first half (0) or the second half of the alphabet (1)?
ANSWER: In the first half (0).
THIRD QUESTION: Of the thirteen letters in the first half of the alphabet, is it in the first seven (0) or the second six (1)?
ANSWER: In the second six (1).
FOURTH QUESTION: In the second six (H, I, J, K, L, M), is it in the first half (0) or the second half (1)?

ANSWER: In the first half (0).
FIFTH QUESTION: Of these letters H, I, J, is it H (0) or is it one of I and J (1)?
ANSWER: It is one of I and J (1).
SIXTH QUESTION: Is it I (0) or J (1)?
ANSWER: It is J (1).

Specifying the letter J is therefore equivalent to the binary message, 001011. But it required not twenty questions but six, and it is in this sense that only six bits are required to specify a given letter. Therefore twenty billion bits are the equivalent of about three billion letters ($2 \times 10^{10}/6 \cong 3 \times 10^9$). If there are approximately six letters in an average word, the information content of a human chromosome corresponds to about five hundred million words ($3 \times 10^9/6 = 5 \times 10^8$). If there are about three hundred words on an ordinary page of printed type, this corresponds to about two million pages ($5 \times 10^8/3 \times 10^2 \cong 2 \times 10^6$). If a typical book contains five hundred such pages, the information content of a single human chromosome corresponds to some four thousand volumes ($2 \times 10^6/5 \times 10^2 = 4 \times 10^3$). It is clear, then, that the sequence of rungs on our DNA ladders represents an enormous library of information. It is equally clear that so rich a library is required to specify as exquisitely constructed and intricately functioning an object as a human being. Simple organisms have less complexity and less to do, and therefore require a smaller amount of genetic information. The Viking landers that put down on Mars in 1976 each had preprogrammed instructions in their computers amounting to a few million bits. Thus Viking had slightly more "genetic information" than a bacterium, but significantly less than an alga.

The chart on page 22 also shows the minimum amount of genetic information in the DNA of various taxa. The amount shown for mammals is less than for human beings, because most mammals have less genetic information than human beings do. Within certain taxa—for example, the amphibians—the amount of genetic

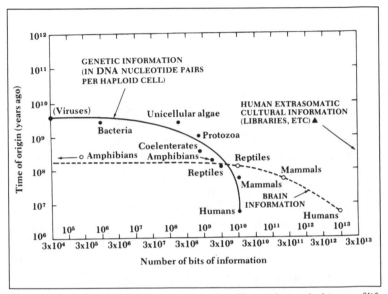

The evolution of information content in genes and brains during the history of life on Earth. The solid curve, which goes with the filled circles, represents the number of bits of information contained in the genes of various taxa, whose rough time of origin in the geological record is also shown. Because of variations in the amount of DNA per cell for certain taxa, only the minimum information content for a given taxon is shown, the data being taken from the work of Britten and Davidson (1969). The dashed curve, which goes with the open circles, is an approximate estimate of the evolution in the amount of information in the brains and nervous systems of these same organisms. The information in the brains of amphibians and still lower animals are off the left edge of the figure. The number of bits of information in the genetic material of viruses is shown, but it is not clear that viruses originated several billions of years ago. It is possible that viruses have evolved more recently, by loss of function, from bacteria or other more elaborate organisms. If the extrasomatic information of human beings were included (libraries, etc.), that point would be far off the lower right edge of the chart.

information varies wildly from species to species, and it is thought that much of this DNA may be redundant or functionless. This is the reason that the chart displays the *minimum* amount of DNA for a given taxon.

We see from the chart that there was a striking improvement in the information content of organisms on Earth some three billion years ago, and a slow increase in the amount of genetic information thereafter. We also see that if more than some tens of billions (several times 10^{10}) of bits of information are necessary for human survival, extragenetic systems will have to provide them: the rate of development of genetic systems is so slow that no source of such additional biological information can be sought in the DNA.

The raw materials of evolution are mutations, inheritable changes in the particular nucleotide sequences that make up the hereditary instructions in the DNA molecule. Mutations are caused by radioactivity in the environment, by cosmic rays from space, or, as often happens, randomly—by spontaneous rearrangements of the nucleotides which statistically must occur every now and then. Chemical bonds spontaneously break. Mutations are also to some extent controlled by the organism itself. Organisms have the ability to repair certain classes of structural damage done to their DNA. There are, for example, molecules which patrol the DNA for damage; when a particularly egregious alteration in the DNA is discovered, it is snipped out by a kind of molecular scissors, and the DNA put right. But such repair is not and must not be perfectly efficient: mutations are required for evolution. A mutation in a DNA molecule within a chromosome of a skin cell in my index finger has no influence on heredity. Fingers are not involved, at least directly, in the propagation of the species. What counts are mutations in the gametes, the eggs and sperm cells, which are the agents of sexual reproduction.

Accidentally useful mutations provide the working material for biological evolution—as, for example, a mutation for melanin in certain moths, which changes their color from white to black. Such moths commonly rest on English birch trees, where their white coloration provides protective camouflage. Under these conditions, the melanin mutation is not an advantage—the dark moths are starkly visible and are eaten by birds; the mutation is selected against. But when the Industrial Revolution began to cover the birch bark with soot, the situation was reversed, and only moths with the melanin mutation survived. Then the mutation is selected for, and, in time, almost all the moths are dark, passing this inheritable change on to future generations. There are still occasional reverse mutations eliminating the melanin adaptation, which would be useful for the moths were English industrial pollution to be controlled. Note that in all this interaction between mutation and natural selection, no moth is making a conscious *effort* to adapt to a changed environment. The process is random and statistical.

Large organisms such as human beings average about one mutation per ten gametes—that is, there is a 10 percent chance that any given sperm or egg cell produced will have a new and inheritable change in the genetic instructions that determine the makeup of the next generation. These mutations occur at random and are almost uniformly harmful—it is rare that a precision machine is improved by a random change in the instructions for making it.

Most of these mutations are also recessive—they do not manifest themselves immediately. Nevertheless, there is already such a high mutation rate that, as several biologists have suggested, a larger complement of genetic DNA would bring about unacceptably high mutation rates: too much would go wrong too often it we had more genes.* If this is true, there must be a practical upper limit to the amount of genetic information that the DNA of larger organisms can accommodate. Thus large and complex organisms, by the mere fact of their existence, have to have substantial resources

---

*To some extent the mutation rate is itself controlled by natural selection, as in our example of a "molecular scissors." But there is likely to be an irreducible minimum mutation rate (1) in order to produce enough genetic experiments for natural selection to operate on, and (2) as an equilibrium between mutations produced, say, by cosmic rays and the most efficient possible cellular repair mechanisms.

of extragenetic information. That information is contained, in all higher animals except Man, almost exclusively in the brain.

What is the information content of the brain? Let us consider two opposite and extreme poles of opinion on brain function. In one view, the brain, or at least its outer layers, the cerebral cortex, is equipotent: any part of it may substitute for any other part, and there is no localization of function. In the other view, the brain is completely hard-wired: specific cognitive functions are localized in particular places in the brain. Computer design suggests that the truth lies somewhere between these two extremes. On the one hand, any nonmystical view of brain function must connect physiology with anatomy; particular brain functions must be tied to particular neural patterns or other brain architecture. On the other hand, to assure accuracy and protect against accident we would expect natural selection to have evolved substantial redundancy in brain function. This is also to be expected from the evolutionary path that it is most likely the brain followed.

The redundancy of memory storage was clearly demonstrated by Karl Lashley, a Harvard psychoneurologist, who surgically removed (extirpated) significant fractions of the cerebral cortex of rats without noticeably affecting their recollection of previously learned behavior on how to run mazes. From such experiments it is clear that the same memory must be localized in many different places in the brain, and we now know that some memories are funneled between the left and right cerebral hemispheres by a conduit called the corpus callosum.

Lashley also reported no apparent change in the general behavior of a rat when significant fractions—say, 10 percent—of its brain were removed. But no one asked the rat its opinion. To investigate this question properly would require a detailed study of rat social, foraging, and predator-evasion behavior. There are many conceivable behavioral changes resulting from such extirpations that might not be immediately obvious to the casual scientist but that might be of considerable significance to the rat—such as the amount of post-extirpation interest an attractive rat of the opposite sex now elicits, or the degree of disinterest now evinced by the presence of a stalking cat.*

It is sometimes argued that cuts or lesions in significant parts of the cerebral cortex in humans—as by bilateral prefrontal lobotomy or by an accident—have little effect on behavior. But some sorts of human behavior are not very apparent from the outside, or even from the inside. There are human perceptions and activities that may occur only rarely, such as creativity. The association of ideas involved in acts—even small ones—of creative genius seems to imply substantial investments of brain resources. These creative acts indeed characterize our entire civilization and mankind as a species. Yet in many people they occur only rarely, and their absence may be missed by neither the brain-damaged subject nor the inquiring physician.

While substantial redundancy in brain function is inevitable, the strong equipotent hypothesis is almost certainly wrong, and most contemporary neurophysiologists have rejected it. On the other hand, a weaker equipotent hypothesis—holding, for example, that memory is a function of the cerebral cortex as a whole—is not so readily dismissable, although it is testable, as we shall see.

There is a popular contention that half or more of the brain is unused. From an evolutionary point of view this would be quite extraordinary: why should it have evolved if it had no function? But actually the statement is made on very little evidence. Again, it is deduced from the finding that many lesions of the brain, generally of the cerebral cortex, have no apparent effect on behavior. This view does not take into account (1) the possibility of redundant function; and (2) the fact that some human behavior is subtle. For

---

*Incidentally, as a test of the influence of animated cartoons on American life, try rereading this paragraph with the word "rat" replaced everywhere by "mouse," and see if your sympathy for the surgically invaded and misunderstood beast suddenly increases.

example, lesions in the right hemisphere of the cerebral cortex may lead to impairments in thought and action, but in the nonverbal realm, which is, by definition, difficult for the patient or the physician to describe.

There is also considerable evidence for localization of brain function. Specific brain sites below the cerebral cortex have been found to be concerned with appetite, balance, thermal regulation, the circulation of the blood, precision movements and breathing. A classic study on higher brain function is the work of the Canadian neurosurgeon, Wilder Penfield, on the electrical stimulation of various parts of the cerebral cortex, generally in attempts to relieve symptoms of a disease such as psychomotor epilepsy. Patients reported a snatch of memory, a smell from the past, a sound or color trace— all elicited by a small electrical current at a particular site in the brain.

In a typical case, a patient might hear an orchestral composition in full detail when current flowed through Penfield's electrode to the patient's cortex, exposed after a craniotomy. If Penfield indicated to the patient—who typically is fully conscious during such procedures—that he was stimulating the cortex when he was not, invariably the patient would report no memory trace at that moment. But when, without notice, a current would flow through the electrode into the cortex, a memory trace would begin or continue. A patient might report a feeling tone, or a sense of familiarity, or a full retrieval of an experience of many years previous playing back in his mind, simultaneously but in no conflict with his awareness of being in an operating room conversing with a physician. While some patients described these flashbacks as "little dreams," they contained none of the characteristic symbolism of dream material. These experiences have been reported almost exclusively by epileptics, and it is possible, although it has by no means been demonstrated, that non-epileptics are, under similar circumstances, subject to comparable perceptual reminiscences.

In one case of electrical stimulation of the occipital lobe, which is concerned with vision, the patient reported seeing a fluttering butterfly of such compelling reality that he stretched out his hand from the operating table to catch it. In an identical experiment performed on an ape, the animal peered intently, as if at an object before him, made a swift catching motion with his right hand, and then examined, in apparent bewilderment, his empty fist.

Painless electrical stimulation of at least some human cerebral cortices elicits cascades of memories of particular events. But removal of the brain tissue in contact with the electrode does not erase the memory. It is difficult to resist the conclusion that at least in humans memories are stored somewhere in the cerebral cortex, waiting for the brain to retrieve them by electrical impulses—which, of course, are ordinarily generated within the brain itself.

If memory is a function of the cerebral cortex as a whole—a kind of dynamic reverberation or electrical standing wave pattern of the constituent parts, rather than stored statically in separate brain components—this would explain the survival of memory after significant brain damage. The evidence, however, points in the other direction: In experiments performed by the American neurophysiologist Ralph Gerard at the University of Michigan, hamsters were taught to run a simple maze and then chilled almost to the freezing point in a refrigerator, a kind of induced hibernation. The temperatures were so low that all detectable electrical activity in the animals' brains ceased. If the dynamic view of memory were true, the experiment should have wiped out all memory of successful maze-running. Instead, after thawing, the hamsters remembered. Memory seems to be localized in specific sites in the brain, and the survival of memories after massive brain lesions must be the result of redundant storage of static memory traces in various locales.

Penfield, extending the findings of previous researchers, also uncovered a remarkable localization of function in the motor cortex. Certain parts of the outer layers of our brain are responsible for sending signals to or receiving signals from specific parts of the body. A version of Penfield's maps of the sensory and motor cortices appear on pages

30 and 31. It reflects in an engaging way the relative importance of various parts of our body. The enormous amount of brain area committed to the fingers—particularly the thumb—and to the mouth and the organs of speech corresponds precisely to what in human physiology, through human behavior, has set us apart from most of the other animals. Our learning and our culture would never have developed without speech; our technology and our monuments would never have evolved without hands. In a way, the map of the motor cortex is an accurate portrait of our humanity.

But the evidence for localization of function is now much stronger even than this. In an elegant set of experiments, David Hubel of Harvard Medical School discovered the existence of networks of particular brain cells that respond selectively to lines perceived by the eye in different orientations. There are cells for horizontal, and cells for vertical, and cells for diagonal, each of which is stimulated only if lines of the appropriate orientation are perceived. At least some beginnings of abstract thought have thereby been traced to the cells of the brain.

The existence of specific brain areas dealing with particular cognitive, sensory or motor functions implies that there need not be any perfect correlation between brain mass and intelligence; some parts of the brain are clearly more important than others. Among the most massive human brains on record are those of Oliver Cromwell, Ivan Turgenev and Lord Byron, all of whom were smart but no Albert Einsteins. Einstein's brain, on the other hand, was not remarkably large. Anatole France, who was brighter than many, had a brain half the size of Byron's. The human baby is born with an exceptionally high ratio of brain mass to body mass (about 12 percent); and the brain, particularly the cerebral cortex, continues to grow rapidly in the first three years of life—the period of most rapid learning. By age six, the mass of the brain is 90 percent of its adult value. The average mass of the brain of contemporary men is about 1,375 grams, almost three pounds. Since the density of the brain, like that of all body tissues, is about that of water (one gram per cubic centimeter), the volume of such a brain is 1,375 cubic centimeters, a little under a liter and a half. (One cubic centimeter is about the volume of an adult human navel.)

But the brain of a contemporary woman is about 150 cubic centimeters smaller. When cultural and child-rearing biases are taken

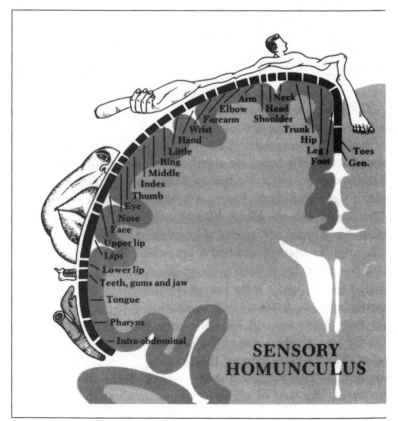

Sensory and motor homunculi, after Penfield. These are two maps of the specialization of function in the cerebral cortex. The distorted mannequins are maps of how much attention is given in the cortex to various parts of the body; the larger

into account, there is no clear evidence of overall differences in intelligence between the sexes. Therefore, brain-mass differences of 150 grams in humans must be unimportant. Comparable differences in brain mass exist among adults of different human races (Orientals, on the average, have slightly larger brains than whites); since no differences in intelligence under similarly controlled

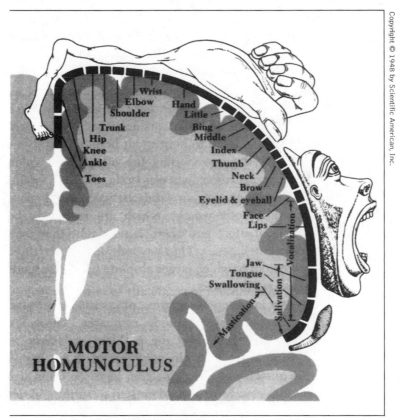

MOTOR
HOMUNCULUS

the body part shown, the more important it is. At left is a map of the somatic sensory area, which receives neural information from the parts of the body shown; at right is a corresponding map for the transmission of impulses from brain to body.

conditions have been demonstrated there, the same conclusion follows. And the gap between the sizes of the brains of Lord Byron (2,200 grams) and Anatole France (1,100 grams) suggests that, in this range, differences of many hundreds of grams may be functionally unimportant.

On the other hand, adult human microcephalics, who are born with tiny brains, have vast losses in cognitive abilities; their typical brain masses are between 450 and 900 grams. A normal newborn child has a typical brain mass of 350 grams; a one-year-old, about 500 grams. It is clear that, as we consider smaller and smaller brain masses, there comes a point where the brain mass is so tiny that its function is severely impaired, compared to normal adult human brain function.

Moreover, there is a statistical correlation between brain mass or size and intelligence in human beings. The relationship is not one-to-one, as the Byron–France comparison clearly shows. We cannot tell a person's intelligence in any given case by measuring his or her brain size. However, as the American evolutionary biologist Leigh van Valen of the University of Chicago has shown, the available data suggest a fairly good correlation, on the average, between brain size and intelligence. Does this mean that brain size in some sense *causes* intelligence? Might it not be, for example, that malnutrition, particularly *in utero* and in infancy, leads to both small brain size and low intelligence, without the one causing the other? Van Valen points out that the correlation between brain size and intelligence is much better than the correlation between intelligence and stature or adult body weight, which are known to be influenced by malnutrition, and there is no doubt that malnutrition can lower intelligence. Thus beyond such effects, there appears to be an extent to which larger absolute brain size tends to produce higher intelligence.

In exploring new intellectual territory, physicists have found it useful to make order-of-magnitude estimates. These are rough calculations that block out the problem and serve as guides for future studies. They do not pretend to be highly accurate. In the question of the connection between brain size and intelligence, it is clearly far beyond present scientific abilities to perform a census of the function of every cubic centimeter of the brain. But might there not be some rough and approximate way in which to connect brain mass with intelligence?

The difference in brain mass between the sexes is of interest in precisely this context, because women are systematically smaller in size and have a lower body mass than men. With less body to control, might not a smaller brain mass be adequate? This suggests that a better measure of intelligence than the absolute value of the mass of a brain is the *ratio* of the mass of the brain to the total mass of the organism.

The chart on page 34 shows the brain masses and body masses of various animals. There is a remarkable separation of fish and reptiles from birds and mammals. For a *given* body mass or weight, mammals have consistently higher brain mass. The brains of mammals are ten to one hundred times more massive than the brains of contemporary reptiles of comparable size. The discrepancy between mammals and dinosaurs is even more striking. These are stunningly large and completely systematic differences. Since we are mammals, we probably have some prejudices about the relative intelligence of mammals and reptiles; but I think the evidence is quite compelling that mammals are indeed systematically much more intelligent than reptiles. (Also shown is an intriguing exception: a small ostrich-like theropod class of dinosaurs from the late Cretaceous Period, whose ratio of brain to body mass places them just within the regional diagram otherwise restricted to large birds and the less intelligent mammals. It would be interesting to know much more about these creatures, which have been studied by Dale Russell, chief of the Palaeontology Division of the National Museums of Canada.) We also see from the chart on page 34 that the primates, a taxon that includes man, are separated, but less systematically, from the rest of the mammals; primate brains are on the average more massive by a factor of about two to twenty than those of nonprimate mammals of the same body mass.

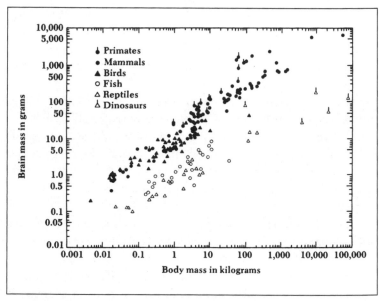

A scatter diagram of brain mass versus body mass for primates, mammals, birds, fish, reptiles, and dinosaurs. The diagram has been adapted from the work of Jerison (1973), with some points added for the dinosaurs and now-extinct members of the family of man.

When we look more closely at this chart, isolating a number of particular animals, we see the results on page 35. Of all the organisms shown, the beast with the largest brain mass for its body weight is a creature called *Homo sapiens*. Next in such a ranking are the dolphins.* Again I do not think it is chauvinistic to conclude from evidence on their behavior that humans and dolphins are at least among the most intelligent organisms on Earth.

*By the criterion of brain mass to body mass, sharks are the smartest of the fishes, which is consistent with their ecological niche—predators have to be brighter than plankton browsers. Both in their increasing ratio of brain to body mass and in the development of coordinating centers in the three principal components of their brains, sharks have evolved in a manner curiously parallel to the evolution of higher vertebrates on the land.

The importance of this ratio of brain to body mass had been realized even by Aristotle. Its principal modern exponent has been Harry Jerison, a neuropsychiatrist at the University of California at Los Angeles. Jerison points out that some exceptions exist to our correlation—e.g., the European pygmy shrew has a brain mass of 100 milligrams in a 4.7 gram body, which gives it a mass ratio in the human range. But we cannot expect the correlation of mass ratio with intelligence to apply to the smallest animals, because the simplest "housekeeping" functions of the brain must require some minimum brain mass.

The brain mass of a mature sperm whale, a close relative of the dolphin, is almost 9,000 grams, six and a half times that of the average man. It is unusual in total brain mass, not (compare with the figure below) in ratio of brain to body weight. Yet the largest

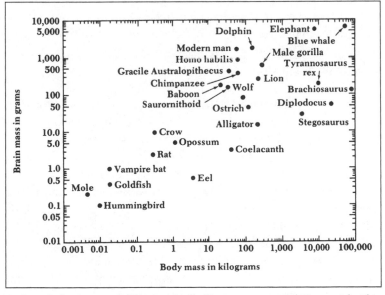

A closer look at some of the points in the diagram on page 34. Saurornithoid is the ostrich-like dinosaur mentioned in the text.

dinosaurs had brain weights about 1 percent that of the sperm whale. What does the whale do with so massive a brain? Are there thoughts, insights, arts, sciences and legends of the sperm whale?

The criterion of brain mass to body mass, which involves no considerations of behavior, appears to provide a very useful index of the relative intelligence of quite different animals. It is what a physicist might describe as an acceptable first approximation. (Note for future reference that the Australopithecines, who were either ancestral to man or at least close collateral relatives, also had a large brain mass for their body weight; this has been determined by making casts of fossil braincases.) I wonder if the unaccountable general appeal of babies and other small mammals—with relatively large heads compared to adults of the same species—derives from our unconscious awareness of the importance of brain to body mass ratios.

The data so far in this discussion suggest that the evolution of mammals from reptiles over two hundred million years ago was accompanied by a major increase in relative brain size and intelligence; and that the evolution of human beings from nonhuman primates a few million years ago was accompanied by an even more striking development of the brain.

The human brain (apart from the cerebellum, which does not seem to be involved in cognitive functions) contains about ten billion switching elements called neurons. (The cerebellum, which lies beneath the cerebral cortex, toward the back of the head, contains roughly another ten billion neurons.) The electrical currents generated by and through the neurons or nerve cells were the means by which the Italian anatomist Luigi Galvani discovered electricity. Galvani had found that electrical impulses could be conducted to the legs of frogs, which dutifully twitched; and the idea became popular that animal motion ("animation") was in its deepest sense caused by electricity. This is at best a partial truth; electrical impulses transmitted along nerve fibers do, through neurochemical intermediaries, initiate such movements as the articulation of limbs, but the impulses are generated in the brain.

Nevertheless, the modern science of electricity and the electrical and electronic industries all trace their origins to eighteenth-century experiments on the electrical stimulation of twitches in frogs.

Only a few decades after Galvani, a group of literary English-persons, immobilized in the Alps by inclement weather, set themselves a competition to write a fictional work of consummate horror. One of them, Mary Wollstonecraft Shelley, penned the now-famous tale of Dr. Frankenstein's monster, who is brought to life by the application of massive electrical currents. Electrical devices have been a mainstay of gothic novels and horror films ever since. The essential idea is Galvani's and is fallacious, but the concept has insinuated itself into many Western languages—as, for example, when I am galvanized into writing this book.

Most neurobiologists believe that the neurons are the active elements in brain function, although there is evidence that some specific memories and other cognitive functions may be contained in particular molecules in the brain, such as RNA or small proteins. For every neuron in the brain there are roughly ten glial cells (from the Greek word for glue), which provide the scaffolding for the neuronal architecture. An average neuron in a human brain has between 1,000 and 10,000 synapses or links with adjacent neurons. (Many spinal-cord neurons seem to have about 10,000 synapses, and the so-called Purkinje cells of the cerebellum may have still more. The number of links for neurons in the cortex is probably less than 10,000.) If each synapse responds by a single yes-or-no answer to an elementary question, as is true of the switching elements in electronic computers, the maximum number of yes/no answers or bits of information that the brain could contain is about $10^{10} \times 10^3 = 10^{13}$, or 10 trillion bits (or 100 trillion = 10^{14} bits if we had used 10^4 synapses per neuron). Some of these synapses must contain the same information as is contained in other synapses; some must be concerned with motor and other noncognitive functions; and some may be merely blank, a buffer waiting for the new day's information to flutter through.

If each human brain had only one synapse—corresponding to a monumental stupidity—we would be capable of only two mental states. If we had two synapses, then $2^2 = 4$ states; three synapses, then $2^3 = 8$ states, and, in general, for N synapses, 2^N states. But the human brain is characterized by some 10^{13} synapses. Thus the number of different states of a human brain is 2 raised to this power—i.e., multiplied by itself ten trillion times. This is an unimaginably large number, far greater, for example, than the total number of elementary particles (electrons and protons) in the entire universe, which is much less than 2 raised to the power 10^3. It is because of this immense number of functionally different configurations of the human brain that no two humans, even identical twins raised together, can ever be really very much alike. These enormous numbers may also explain something of the unpredictability of human behavior and those moments when we surprise even ourselves by what we do. Indeed, in the face of these numbers, the wonder is that there are any regularities at all in human behavior. The answer must be that all possible brain states are by no means occupied; there must be an enormous number of mental configurations that have never been entered or even glimpsed by any human being in the history of mankind. From this perspective, each human being is truly rare and different and the sanctity of individual human lives is a plausible ethical consequence.

In recent years it has become clear that there are electrical microcircuits in the brain. In these microcircuits the constituent neurons are capable of a much wider range of responses than the simple "yes" or "no" of the switching elements in electronic computers. The microcircuits are very small in size (typical dimensions are about 1/10,000 of a centimeter) and thus able to process data very rapidly. They respond to about 1/100th of the voltage necessary to stimulate ordinary neurons, and are therefore capable of much finer and subtler responses. Such microcircuits seem to increase in abundance in a manner consistent with our usual notions about the complexity of an animal, reaching their greatest proliferation in both absolute and relative terms in human beings. They also develop late in human embryology. The existence of such microcircuits suggests that intelligence may be the result not only of high brain-to-body-mass ratios but also of an abundance of specialized switching elements in the brain. Microcircuits make the number of possible brain states even greater than we calculated in the previous paragraph, and so enhance still farther the astonishing uniqueness of the individual human brain.

We can approach the question of the information content of the human brain in a quite different way—introspectively. Try to imagine some visual memory, say from your childhood. Look at it very closely in your mind's eye. Imagine it is composed of a set of fine dots like a newspaper wirephoto. Each dot has a certain color and brightness. You must now ask how many bits of information are necessary to characterize the color and brightness of each dot; how many dots make up the recalled picture; and how long it takes to recall all the details of the picture in the eye of the mind. In this retrospective, you focus on a very small part of the picture at any one time; your field of view is quite limited. When you put in all of these numbers, you come out with a rate of information processing by the brain, in bits per second. When I do such a calculation, I come out with a peak processing rate of about 5,000 bits per second.*

*Horizon to horizon comprises an angle of 180 degrees in a flat place. The moon is 0.5 degrees in diameter. I know I can see detail on it, perhaps twelve picture elements across. Thus my eye can resolve about 0.5/12 = 0.04 degrees. Anything smaller than this is too small for me to see. The instantaneous field of view in my mind's eye, as well as in my real eye, seems to be something like 2 degrees on a side. Thus the little square picture I can see at any given moment contains about $(2/0.04)^2 = 2,500$ picture elements, corresponding to the wirephoto dots. To characterize all possible shades of gray and colors of such dots requires about 20 bits per picture element. Thus a description of my little picture requires $2,500 \times 20$ or about 50,000 bits. But the act of scanning the picture takes about 10 seconds, and thus my sensory data processing rate is probably not much larger than $50,000/10 = 5,000$ bits per second. For comparison, the Viking lander cameras, which also have a 0.04 degree resolution, have only six bits per picture element to characterize brightness, and can transmit these directly to Earth by radio at 500 bits per second. The neurons of the brain generate about 25 watts of power, barely enough to turn on a small incandescent light. The Viking lander transmits radio messages and performs all its other functions with a total power of about 50 watts.

Most commonly such visual recollections concentrate on the edges of forms and sharp changes from bright to dark, and not on the configuration of areas of largely neutral brightness. The frog, for example, sees with a very strong bias towards brightness gradients. However, there is considerable evidence that detailed memory of interiors and not just edges of forms is reasonably common. Perhaps the most striking case is an experiment with humans on stereo reconstruction of a three-dimensional image, using a pattern recalled for one eye and a pattern being viewed for the other. The fusion of images in this anaglyph requires a memory of 10,000 picture elements.

But I am not recollecting visual images all my waking hours, nor am I continuously subjecting people and objects to intense and careful scrutiny. I am doing that perhaps a small percent of the time. My other information channels—auditory, tactile, olfactory and gustatory—are involved with much lower transfer rates. I conclude that the average rate of data processing by my brain is about $(5000/50)$ $= 100$ bits per second. Over sixty years, that corresponds to 2×10^{11} or 200 billion total bits committed to visual and other memory if I have perfect recall. This is less than, but not unreasonably less than, the number of synapses or neural connections (since the brain has more to do than just remember) and suggests that neurons are indeed the main switching elements in brain function.

A remarkable series of experiments on brain changes during learning has been performed by the American psychologist Mark Rosenzweig and his colleagues at the University of California at Berkeley. They maintained two different populations of laboratory rats—one in a dull, repetitive, impoverished environment; the other in a variegated, lively, enriched environment. The latter group displayed a striking increase in the mass and thickness of the cerebral cortex, as well as accompanying changes in brain chemistry. These increases occurred in mature as well as in young animals. Such experiments demonstrate that physiological changes accompany intellectual experience and show how plasticity can be controlled anatomically. Since a more massive cerebral cortex may make future learning easier, the importance of enriched environments in childhood is clearly drawn.

This would mean that new learning corresponds to the generation of new synapses or the activation of moribund old ones, and some preliminary evidence consistent with this view has been obtained by the American neuroanatomist William Greenough of the University of Illinois and his co-workers. They have found that after several weeks of learning new tasks in laboratory contexts, rats develop the kind of new neural branches in their cortices that form synapses. Other rats, handled similarly but given no comparable education, exhibit no such neuroanatomical novelties. The construction of new synapses requires the synthesis of protein and RNA molecules. There is a great deal of evidence showing that these molecules are produced in the brain during learning, and some scientists have suggested that the learning is contained within brain proteins or RNA. But it seems more likely that the new information is contained in the neurons, which are in turn constructed of proteins and RNA.

How densely packed is the information stored in the brain? A typical information density during the operation of a modern computer is about a million bits per cubic centimeter. This is the total information content of the computer, divided by its volume. The human brain contains, as we have said, about 10^{13} bits in a little more than 10^3 cubic centimeters, for an information content of $10^{13}/10^3 = 10^{10}$, about ten billion bits per cubic centimeter; the brain is therefore ten thousand times more densely packed with information than is a computer, although the computer is much larger. Put another way, a modern computer able to process the information in the human brain would have to be about ten thousand times larger in volume than the human brain. On the other hand, modern electronic computers are capable of processing information at a rate of 10^{16} to 10^{17} bits per second, compared to a peak rate ten billion times slower in the brain. The brain must be extraordinarily cleverly packaged and "wired," with such a small total information content and so low a processing rate, to be able to do so many significant tasks so much better than the best computer.

The number of neurons in an animal brain does not double as the brain volume itself doubles. It increases more slowly. A human brain with a volume of about 1,375 cubic centimeters contains, as we have said, apart from the cerebellum about ten billion neurons and some ten trillion bits. In a laboratory at the National Institute of Mental Health near Bethesda, Maryland, I recently held in my hand a rabbit brain. It had a volume of perhaps thirty cubic centimeters, the size of an average radish, corresponding to a few hundred million neurons and some hundred billion bits—which controlled, among other things, the munching of lettuce, the twitchings of noses, and the sexual dalliances of grownup rabbits.

Since animal taxa such as mammals, reptiles or amphibians contain members with very different brain sizes, we cannot give a reliable estimate of the number of neurons in the brain of a typical representative of each taxon. But we can estimate average values which I have done in the chart on page 22. The rough estimates there show that a human being has about a hundred times more bits of information in his brain than a rabbit does. I do not know that it means very much to say that a human being is a hundred times smarter than a rabbit, but I am not certain that it is a ridiculous contention. (It does not, of course, follow that a hundred rabbits are as smart as one human being.)

We are now in a position to compare the gradual increase through evolutionary time of both the amount of information contained in the genetic material and the amount of information contained in the brains of organisms. The two curves cross (page 22) at a time corresponding to a few hundred million years ago and at an information content corresponding to a few billion bits. Somewhere in the steaming jungles of the Carboniferous Period there emerged an organism that for the first time in the history of the world had more information in its brains than in its genes. It was an early reptile which, were we to come upon it in these sophisticated times, we would probably not describe as exceptionally intelligent. But its brain was a symbolic turning point in the history of life. The two subsequent bursts of brain evolution, accompanying the emergence of mammals and the advent of manlike primates, were still more important advances in the evolution of intelligence. Much of the history of life since the Carboniferous Period can be described as the gradual (and certainly incomplete) dominance of brains over genes.

When shall we three meet again …?

WM. SHAKESPEARE
Macbeth

CHAPTER III

THE BRAIN
AND THE CHARIOT

THE BRAIN of a fish isn't much. A fish has a notochord or spinal cord, which it shares with even humbler invertebrates. A primitive fish also has a little swelling at the front end of the spinal cord, which is its brain. In higher fish the swelling is further developed but still weighs no more than a gram or two. That swelling corresponds in higher animals to the hindbrain or brainstem and the midbrain. The brains of modern fish are chiefly midbrain, with a tiny forebrain; in modern amphibians and reptiles, it is the other way around (see figure on page 46). And yet fossil endocasts of the earliest known vertebrates show that the principal divisions of the modern brain (hindbrain, midbrain and forebrain, for example) were already established. Five hundred million years ago, swimming in the primeval seas, there were fishy creatures called ostracoderms and placoderms, whose brains had recognizably the same major divisions as ours. But the relative size and importance of these components, and even their early functions, were certainly very different from today. One of the most engaging views of the subsequent evolution of the brain

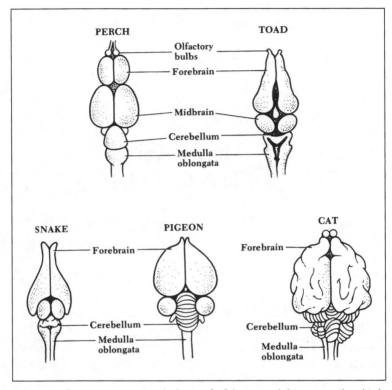

Schematic diagrams comparing the brain of a fish, an amphibian, a reptile, a bird, and a mammal. The cerebellum and medulla oblongata are parts of the hindbrain.

is a story of the successive accretion and specialization of three further layers surmounting the spinal cord, hindbrain and midbrain. After each evolutionary step, the older portions of the brain still exist and must still be accommodated. But a new layer with new functions has been added.

The principal contemporary exponent of this view is Paul MacLean, chief of the Laboratory of Brain Evolution and Behavior of the National Institute of Mental Health. One hallmark of MacLean's work is that it encompasses many different animals, ranging from lizards to squirrel monkeys. Another is that he and his colleagues have studied carefully the social and other behavior of these animals to improve their prospects of discovering what part of the brain controls what sort of behavior.

Squirrel monkeys with "gothic" facial markings have a kind of ritual or display which they perform when greeting one another. The males bare their teeth, rattle the bars of their cage, utter a high-pitched squeak, which is possibly terrifying to squirrel monkeys, and lift their legs to exhibit an erect penis. While such behavior would border on impoliteness at many contemporary human social gatherings, it is a fairly elaborate act and serves to maintain dominance hierarchies in squirrel-monkey communities.

MacLean has found that a lesion in one small part of a squirrel monkey's brain will prevent this display while leaving a wide range of other behavior intact, including sexual and combative behavior. The part that is involved is in the oldest part of the forebrain, a part that humans as well as other primates share with our mammalian and reptilian ancestors. In nonprimate mammals and in reptiles, comparable ritualized behavior seems to be controlled in the same part of the brain, and lesions in this reptilian component can impair other automatic types of behavior besides ritual—for example, walking or running.

The connection between sexual display and position in a dominance hierarchy can be found frequently among the primates. Among Japanese macaques, social class is maintained and reinforced by daily mounting: males of lower caste adopt the characteristic submissive sexual posture of the female in oestrus and are briefly and ceremonially mounted by higher-caste males. These mountings are both common and perfunctory. They seem to have little sexual content but rather serve as easily understood symbols of who is who in a complex society.

In one study of the behavior of the squirrel monkey, Caspar, the dominant animal in the colony and by far the most active displayer, was never seen to copulate, although he accounted for two-thirds of the genital display in the colony—most of it directed toward other adult males. The fact that Caspar was highly motivated to establish dominance but insignificantly motivated toward sex suggests that while these two functions may involve identical organ systems, they are quite separate. The scientists studying this colony concluded: "Genital display is therefore considered the most effective social signal with respect to group hierarchy. It is ritualized and seems to acquire the meaning, 'I am the master.' It is most probably derived from sexual activity, but it is used for social communication and separated from reproductive activity. In other words, genital display is a ritual derived from sexual behavior but serving social and not reproductive purposes."

In a television interview in 1976, a professional football player was asked by the talk-show host if it was embarrassing for football players to be together in the locker room with no clothes on. His immediate response: "We strut! No embarrassment at all. It's as if we're saying to each other, 'Let's see what you got, man!'—except for a few, like the specialty team members and the water boy."

The behavioral as well as neuroanatomical connections between sex, aggression and dominance are borne out in a variety of studies. The mating rituals of great cats and many other animals are barely distinguishable, in their early stages, from fighting. It is a commonplace that domestic cats sometimes purr loudly and perversely while their claws are slowly raking over upholstery or lightly clad human skin. The use of sex to establish and maintain dominance is sometimes evident in human heterosexual and homosexual practices (although it is not, of course, the only element in such practices), as well as in many "obscene" utterances. Consider the peculiar circumstance that the most common two-word verbal aggression in English, and in many other languages, refers to an act of surpassing physical pleasure; the English form probably comes from a Germanic and Middle Dutch verb *fokken*, meaning "to strike." This otherwise puzzling usage can be understood as a verbal equivalent of macaque symbolic language, with the initial word "I" unstated but understood by both parties. It and many similar expressions seem to be human ceremonial mountings. As we will see, such behavior probably extends much farther back than the monkeys, back through hundreds of millions of years of geological time.

From experiments such as those with squirrel monkeys, MacLean has developed a captivating model of brain structure and evolution that he calls the triune brain. "We are obliged," he says, "to look at ourselves and the world through the eyes of three quite different mentalities," two of which lack the power of speech. The human brain, MacLean holds, "amounts to three interconnected biological computers," each with "its own special intelligence, its own subjectivity, its own sense of time and space, its own memory, motor, and other functions." Each brain corresponds to a separate major evolutionary step. The three brains are said to be distinguished neuroanatomically and functionally, and contain strikingly different distributions of the neurochemicals dopamine and cholinesterase.

At the most ancient part of the human brain lies the spinal cord; the medulla and pons, which comprise the hindbrain; and the midbrain. This combination of spinal cord, hindbrain and midbrain MacLean calls the neural chassis. It contains the basic neural machinery for reproduction and self-preservation, including regulation of the heart, blood circulation and respiration. In a fish or an amphibian it is almost all the brain there is. But a reptile or higher animal deprived of its forebrain is, according to MacLean, "as motionless and aimless as an idling vehicle without a driver."

Indeed, *grand mal* epilepsy can, I think, be described as a disease in which the cognitive drivers are all turned off because of a kind of electrical storm in the brain, and the victim is left momentarily with nothing operative but his neural chassis. This is a profound impairment, temporarily regressing the victim back several hundreds of millions of years. The ancient Greeks, whose name for the disease we still use, recognized its profound character and called it the disease inflicted by the gods.

MacLean has distinguished three sorts of drivers of the neural chassis. The most ancient of them surrounds the midbrain (and is made up mostly of what neuroanatomists call the olfactostriatum, the corpus striatum, and the globus pallidus). We share it with the other mammals and the reptiles. It probably evolved several hundred million years ago. MacLean calls it the reptilian or R-complex. Surrounding the R-complex is the limbic system, so called because it borders on the underlying brain. (Our arms and legs are called limbs because they are peripheral to the rest of the body.) We share the limbic system with the other mammals but not, in its full elaboration, with the reptiles. It probably evolved more than one hundred

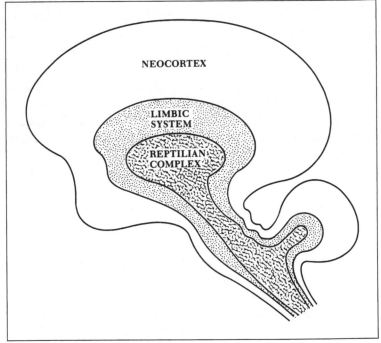

A highly schematic representation of the reptilian complex, limbic system, and neocortex in the human brain, after MacLean.

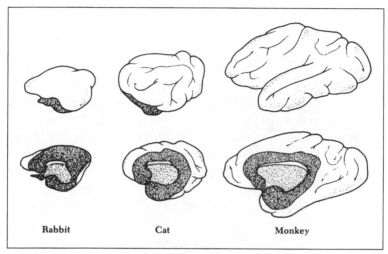

Rabbit   **Cat**   **Monkey**

Schematic views from the top and from the side of the rabbit, cat, and monkey brains. The dark stippled area is the limbic system, seen most easily in the side views. The white furrowed regions represent the neocortex, visible most readily in the top views.

and fifty million years ago. Finally, surmounting the rest of the brain, and clearly the most recent evolutionary accretion, is the neocortex. Like the higher mammals and the other primates, humans have a relatively massive neocortex. It becomes progressively more developed in the more advanced mammals. The most elaborately developed neocortex is ours (and the dolphins' and whales'). It probably evolved several tens of millions of years ago, but its development accelerated greatly a few million years ago when humans emerged. A schematic representation of this picture of the human brain is shown opposite, and a comparison of the limbic system with the neocortex in three contemporary mammals is shown above. The concept of the triune brain is in remarkable accord with the conclusions, drawn independently from studies of brain to body mass ratios in the previous chapter, that the emergence of mammals and of primates (especially humans) was accompanied by major bursts in brain evolution.

It is very difficult to evolve by altering the deep fabric of life; any change there is likely to be lethal. But fundamental change can be accomplished by the addition of new systems on top of old ones. This is reminiscent of a doctrine which was called recapitulation by Ernst Haeckel, a nineteenth-century German anatomist, and which has gone through various cycles of scholarly acceptance and rejection. Haeckel held that in its embryological development, an animal tends to repeat or recapitulate the sequence that its ancestors followed during their evolution. And indeed in human intrauterine development we run through stages very much like fish, reptiles and nonprimate mammals before we become recognizably human. The fish stage even has gill slits, which are absolutely useless for the embryo who is nourished via the umbilical cord, but a necessity for human embryology: since gills were vital to our ancestors, we run through a gill stage in becoming human. The brain of a human fetus also develops from the inside out, and, roughly speaking, runs through the sequence: neural chassis, R-complex, limbic system and neocortex (see the figure on the embryology of the human brain on page 186).

The reason for recapitulation may be understood as follows: Natural selection operates only on individuals, not on species and not very much on eggs or fetuses. Thus the latest evolutionary change appears postpartum. The fetus may have characteristics, like the gill slits in mammals, that are entirely maladaptive after birth, but as long as they cause no serious problems for the fetus and are lost before birth, they can be retained. Our gill slits are vestiges not of ancient fish but of ancient fish embryos. Many new organ systems develop not by the addition and preservation but by the modification of older systems, as, for example, the modification of fins to legs, and legs to flippers or wings; or feet to hands to feet; or sebaceous glands to mammary glands; or gill arches to ear bones; or shark scales to shark teeth. Thus evolution by addition and the functional preservation of the preexisting structure must occur for one of two reasons— either the old function is required as well as the new one, or there is no way of bypassing the old system that is consistent with survival.

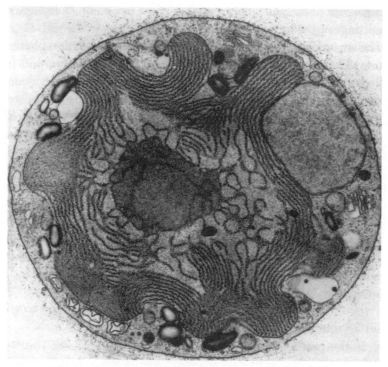

A photograph taken with an electron microscope of a small plant called a red alga. Its scientific name is *Porphyridium cruentum.* The chloroplast, this organism's photosynthetic factory, almost fills the entire cell. The photograph is magnified 23,000 times and was taken by Dr. Elizabeth Gantt of the Smithsonian Institution's Radiation Biology Laboratory.

There are many other examples in nature of this sort of evolutionary development. To take an almost random case, consider why plants are green. Green-plant photosynthesis utilizes light in the red and the violet parts of the solar spectrum to break down water, build up carbohydrates and do other planty things. But the sun gives off more light in the yellow and the green part of the spectrum than in the red or violet. Plants with chlorophyll as their only photosynthetic pigment are rejecting light where it is most plentiful.

Many plants seem belatedly to have "noticed" this and have made appropriate adaptations. Other pigments, which reflect red light and absorb yellow and green light, such as carotenoids and phycobilins, have evolved. Well and good. But have those plants with new photosynthetic pigments abandoned chlorophyll? They have not. The figure on page 53 shows the photosynthetic factory of a red alga. The striations contain the chlorophyll, and the little spheres nestling against these striations contain the phycobilins, which make a red alga red. Conservatively, these plants pass along the energy they acquire from green and yellow sunlight to the chlorophyll pigment that, even though it has not absorbed the light, is still instrumental in bridging the gap between light and chemistry in all plant photosynthesis. Nature could not rip out the chlorophyll and replace it with better pigments; the chlorophyll is woven too deeply into the fabric of life. Plants with accessory pigments are surely different. They are more efficient. But there, still working, although with diminished responsibilities, at the core of the photosynthetic process is chlorophyll. The evolution of the brain has, I think, proceeded analogously. The deep and ancient parts are functioning still.

1 THE R-COMPLEX

If the preceding view is correct, we should expect the R-complex in the human brain to be in some sense performing dinosaur functions still; and the limbic cortex to be thinking the thoughts of pumas and ground sloths. Without a doubt, each new step in brain evolution is accompanied by changes in the physiology of the preexisting components of the brain. The evolution of the R-complex must have seen changes in the midbrain, and so on. What is more, we know that the control of many functions is shared in different components of the brain. But at the same time it would be astonishing if the brain components beneath the neocortex were not to a significant extent still performing as they did in our remote ancestors.

MacLean has shown that the R-complex plays an important role in aggressive behavior, territoriality, ritual and the establishment of social hierarchies. Despite occasional welcome exceptions, this seems to me to characterize a great deal of modern human bureaucratic and political behavior. I do not mean that the neocortex is not functioning at all in an American political convention or a meeting of the Supreme Soviet; after all, a great deal of the communication at such rituals is verbal and therefore neocortical. But it is striking how much of our actual behavior—as distinguished from what we say and think about it—can be described in reptilian terms. We speak commonly of a "cold-blooded" killer. Machiavelli's advice to his Prince was "knowingly to adopt the beast."

In an interesting partial anticipation of these ideas, the American philosopher Susanne Langer wrote: "Human life is shot through and through with ritual, as it is also with animalian practices. It is an intricate fabric of reason and rite, of knowledge and religion, prose and poetry, fact and dream.... Ritual, like art, is essentially the active termination of a symbolic transformation of experience. It is born in the cortex, not in the 'old brain'; but it is born of an *elementary need* of that organ, once the organ has grown to human estate." Except for the fact that the R-complex *is* in the "old brain," this seems to be right on target.

I want to be very clear about the social implications of the contention that reptilian brains influence human actions. If bureaucratic behavior is controlled at its core by the R-complex, does this mean there is no hope for the human future? In human beings, the neocortex represents about 85 percent of the brain, which is surely some index of its importance compared to the brainstem, R-complex and limbic system. Neuroanatomy, political history, and introspection all offer evidence that human beings are quite capable of resisting the urge to surrender to every impulse of the reptilian brain. There is no way, for example, in which the Bill of Rights of the U.S. Constitution could have been recorded, much less conceived, by the R-complex. It is precisely our plasticity, our long childhood, that prevents a slavish adherence to genetically preprogrammed behavior in human beings more than in any other species. But if the triune brain is an accurate model of how human beings function, it does no good whatever to ignore the reptilian component of human nature, particularly our ritualistic and hierarchical behavior. On the contrary, the model may help us to understand what human beings are about. (I wonder, for example, whether the ritual aspects of many psychotic illnesses—e.g., hebephrenic schizophrenia—could be the result of hyperactivity of some center in the R-complex, or of a failure of some neocortical site whose function is to repress or override the R-complex. I also wonder whether the frequent ritualistic behavior in young children is a consequence of the still-incomplete development of their neocortices.)

In a curiously apt passage, G. K. Chesterton wrote: "You can free things from alien or accidental laws, but not from the laws of their own nature.... Do not go about ... encouraging triangles to break out of the prison of their three sides. If a triangle breaks out of its three sides, its life comes to a lamentable end." But not all triangles are equilateral. Some substantial adjustment of the relative role of each component of the triune brain is well within our powers.

2 THE LIMBIC SYSTEM

The limbic system appears to generate strong or particularly vivid emotions. This immediately suggests an additional perspective on the reptilian mind: it is not characterized by powerful passions and wrenching contradictions but rather by a dutiful and stolid acquiescence to whatever behavior its genes and brains dictate.

Electrical discharges in the limbic system sometimes result in symptoms similar to those of psychoses or those produced by psychedelic or hallucinogenic drugs. In fact, the sites of action of many psychotropic drugs are in the limbic system. Perhaps it controls exhilaration and awe and a variety of subtle emotions that we sometimes think of as uniquely human.

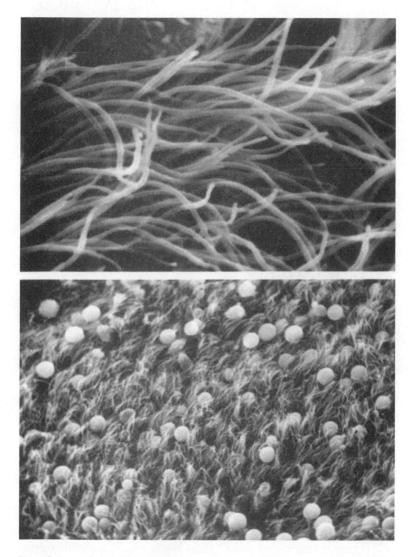

Two photographs taken with an electron microscope within the third ventricle of the brain by Richard Steger of Wayne State University. Tiny waving hairs or cilia can be seen transporting small spherical brain proteins—like a crowd passing large beach balls overhead.

The "master gland," the pituitary, which influences other glands and dominates the human endocrine system, is an intimate part of the limbic region. The mood-altering qualities of endocrine imbalances give us an important hint about the connection of the limbic system with states of mind. There is a small almond-shaped inclusion in the limbic system called the amygdala which is deeply involved in both aggression and fear. Electrical stimulation of the amygdala in placid domestic animals can rouse them to almost unbelievable states of fear or frenzy. In one case, a house cat cowered in terror when presented with a small white mouse. On the other hand, naturally ferocious animals, such as the lynx, become docile and tolerate being petted and handled when their amygdalas are extirpated. Malfunctions in the limbic system can produce rage, fear or sentimentality that have no apparent cause. Natural hyperstimulation may produce the same results: those suffering from such a malady find their feelings inexplicable and inappropriate; they may be considered mad.

At least some of the emotion-determining role of such limbic endocrine systems as the pituitary, amygdala, and hypothalamus is provided by small hormonal proteins which they exude, and which affect other areas of the brain. Perhaps the best-known is the pituitary protein, ACTH (adrenocorticotropic hormone), which can affect such diverse mental functions as visual retention, anxiety and attention span. Some small hypothalamic proteins have been identified tentatively in the third ventricle of the brain, which connects the hypothalamus with the thalamus, a region also within the limbic system. The stunning pictures on page 57, taken with an electron microscope, show two close-ups of action in the third ventricle. The diagram on page 64 may help clarify some of the brain anatomy just described.

There are reasons to think that the beginnings of altruistic behavior are in the limbic system. Indeed, with rare exceptions (chiefly the social insects), mammals and birds are the only organisms to devote substantial attention to the care of their young—an

An impression of the possible form of the Mesozoic reptile *Lycaenops* by John Germann. Such mammal-like creatures were perhaps among the first to experience a substantial evolution of the limbic system.

evolutionary development that, through the long period of plasticity which it permits, takes advantage of the large information-processing capability of the mammalian and primate brains. Love seems to be an invention of the mammals.*

Much in animal behavior substantiates the notion that strong emotions evolved chiefly in mammals and to a lesser extent in birds. The attachment of domestic animals to humans is, I think, beyond

---

*This rule on the relative parental concern of mammals and reptiles is, however, by no means without exceptions. Nile crocodile mothers carefully put their fresh hatchlings in their mouths and carry them to the comparative safety of the river waters; while Serengeti male lions will, upon newly dominating a pride, destroy all the resident cubs. But on the whole, mammals show a strikingly greater degree of parental care than do reptiles. The distinction may have been even more striking one hundred million years ago.

question. The apparently sorrowful behavior of many mammalian mothers when their young are removed is well-known. One wonders just how far such emotions go. Do horses on occasion have glimmerings of patriotic fervor? Do dogs feel for humans something akin to religious ecstasy? What other strong or subtle emotions are felt by animals that do not communicate with us?

The oldest part of the limbic system is the olfactory cortex, which is related to smell, the haunting emotional quality of which is familiar to most humans. A major component of our ability to remember and recall is localized in the hippocampus, a structure within the limbic system. The connection is clearly shown by the profound memory impairment that results from lesions of the hippocampus. In one famous case, H. M., a patient with a long history of seizures and convulsions, was subjected to a bilateral extirpation of the entire region about the hippocampus in a successful attempt to reduce their frequency and severity. He immediately became amnesic. He retained good perceptual skills, was able to learn new motor skills and experienced some perceptual learning but essentially forgot everything more than a few hours old. His own comment was "Every day is alone in itself—whatever enjoyment I've had and whatever sorrow I've had." He described his life as a continuous extension of the feeling of disorientation many of us have upon awakening from a dream, when we have great difficulty remembering what has just happened. Remarkably enough, despite this severe impairment, his IQ improved after his hippocampectomy. He apparently could detect smells but had difficulty identifying by name the source of the smell. He also exhibited an apparent total disinterest in sexual activity.

In another case, a young American airman was injured in a mock duel with another serviceman, when a miniature fencing foil was plunged into his right nostril, puncturing a small part of the limbic system immediately above. This resulted in a severe impairment of memory, similar to but not so severe as H. M.'s; a wide range of his perceptual and intellectual abilities was unaffected. His memory impairment was particularly noticeable with verbal material. In addition, the accident seems to have rendered him both impotent and unresponsive to pain. He once walked barefoot on the sun-heated metal deck of a cruise ship, without realizing that his feet were being badly burned until his fellow passengers complained of the uncomfortable odor of charring flesh. On his own, he was aware of neither the pain nor the smell.

From such cases, it seems apparent that so complex a mammalian activity as sex is controlled simultaneously by all three components of the triune brain—the R-complex, the limbic system and the neo-cortex. (We have already mentioned the involvement of the R-complex and the limbic system in sexual activity. Evidence for involvement of the neocortex can be easily obtained by introspection.)

One segment of the old limbic system is devoted to oral and gustatory functions; another, to sexual functions. The connection of sex with smell is very ancient, and is highly developed in insects—a circumstance that offers insights into both the importance and the disadvantages of reliance on smell in our remote ancestors.

I once witnessed an experiment in which the head of a green bottle fly was connected by a very thin wire to an oscilloscope that displayed, in a kind of graph, any electrical impulses produced by the fly's olfactory system. (The fly's head had only recently been severed from its body—in order to gain access to the olfactory apparatus—and was still in many respects functional.*) The experimenters wafted a wide variety of odors in front of it, including obnoxious and irritating gases such as ammonia, with no discernible effects. The line traced out on the oscilloscope screen was absolutely flat and horizontal. Then a tiny quantity of the sex attractant released by

---

*The heads and bodies of arthropods can briefly function without each other very nicely. A female praying mantis will often respond to earnest courting by decapitating her suitor. While this would be viewed as unsociable among humans, it is not so among insects: extirpation of the brain removes sexual inhibitions and encourages what is left of the male to mate. Afterwards, the female completes her celebratory repast, dining, of course, alone. Perhaps this represents a cautionary lesson against excessive sexual repression.

the female of the species was waved before the severed head, and an enormous vertical spike obligingly appeared on the oscilloscope screen. The bottle fly could smell almost nothing except the female sex attractant. But that molecule he could smell exceedingly well.

Such olfactory specialization is quite common in insects. The male silkworm moth is able to detect the female's sex attractant molecule if only about forty molecules per second reach its feathery antennae. A single female silkworm moth need release only a hundredth of a microgram of sex attractant per second to attract every male silkworm in a volume of about a cubic mile. That is why there are silkworms.

Perhaps the most curious exploitation of the reliance on smell to find a mate and continue the species is found in a South African beetle, which burrows into the ground during the winter. In the spring, as the ground thaws, the beetles emerge, but the male beetles groggily disinter themselves a few weeks before the females do. In this same region of South Africa, a species of orchid has evolved which gives off an aroma identical to the sex attractant of the female beetle. In fact, orchid and beetle evolution have produced essentially the same molecule. The male beetles turn out to be exceedingly nearsighted; and the orchids have evolved a configuration of their petals that, to a myopic beetle, resembles the female in a receptive sexual posture. The male beetles enjoy several weeks of orgiastic ecstasy among the orchids, and when eventually the females emerge from the ground, we can imagine a great deal of wounded pride and righteous indignation. Meanwhile the orchids have been successfully cross-pollinated by the amorous male beetles, who, now properly abashed, do their best to continue the beetle species; and both organisms survive. (Incidentally, it is in the interest of the orchids not to be too consummately attractive; if the beetles fail to reproduce themselves, the orchids are in trouble.) We thus discover one limitation to purely olfactory sexual stimuli. Another is that since every female beetle produces the same sex attractant, it is not easy for a male beetle to fall in love with the lady insect of his heart's desire. While male insects may display themselves to attract a female, or—as with stag beetles—engage in mandible-to-mandible combat with the female as the prize, the central role of the female sex attractant in mating seems to reduce the extent of sexual selection among the insects.

Other methods of finding a mate have been developed in reptiles, birds and mammals. But the connection of sex with smell is still apparent neuroanatomically in higher animals as well as anecdotally in human experience. I sometimes wonder if deodorants, particularly "feminine" deodorants, are an attempt to disguise sexual stimuli and keep our minds on something else.

3 THE NEOCORTEX

Even in fish, lesions of the forebrain destroy the traits of initiative and caution. In higher animals these traits, much elaborated, seem localized in the neocortex, the site of many of the characteristic human cognitive functions. It is frequently discussed in terms of four major regions or lobes: the frontal, parietal, temporal and occipital lobes. Early neurophysiologists held that the neocortex was primarily connected only to other places in the neocortex, but it is now known that there are many neural connections with the subcortical brain. It is, however, by no means clear that the neocortical subdivisions are actually functional units. Each certainly has many quite different functions, and some functions may be shared among or between lobes. Among other functions, the frontal lobes seem to be connected with deliberation and the regulation of action; the parietal lobes, with spatial perception and the exchange of information between the brain and the rest of the body; the temporal lobes, with a variety of complex perceptual tasks; and the occipital lobes, with vision, the dominant sense in humans and other primates.

For many decades the prevailing view of neurophysiologists was that the frontal lobes, behind the forehead, are the sites of anticipation and planning for the future, both characteristically human

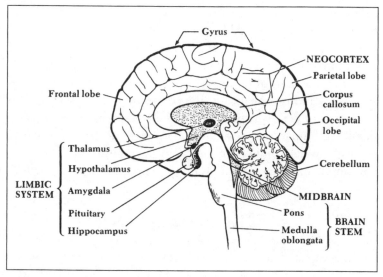

A schematic diagram of a side view of the human brain, dominated by the neo-cortex, with a smaller limbic system and brainstem or hindbrain. The R-complex is not shown.

functions. But more recent work has shown that the situation is not so simple. A large number of cases of frontal lesions—largely suffered in warfare and as gunshot wounds—have been investigated by the American neurophysiologist Hans-Lukas Teuber of the Massachusetts Institute of Technology. He found that many frontal-lobe lesions have almost no obvious effects on behavior; however, in severe pathology of the frontal lobes "the patient is not altogether devoid of capacity to anticipate a course of events, but cannot picture himself in relation to those events as a potential agent." Teuber emphasized the fact that the frontal lobe may be involved in motor as well as cognitive anticipation, particularly in estimating what the effect of voluntary movements will be. The frontal lobes also seem to be implicated in the connection between vision and erect bipedal posture.

Thus the frontal lobes may be involved with peculiarly human functions in two different ways. If they control anticipation of the future, they must also be the sites of concern, the locales of worry. This is why transection of the frontal lobes reduces anxiety. But prefrontal lobotomy must also greatly reduce the patient's capacity to be human. The price we pay for anticipation of the future is anxiety about it. Foretelling disaster is probably not much fun; Pollyanna was much happier than Cassandra. But the Cassandric components of our nature are necessary for survival. The doctrines for regulating the future that they produced are the origins of ethics, magic, science and legal codes. The benefit of foreseeing catastrophe is the ability to take steps to avoid it, sacrificing short-term for long-term benefits. A society that is, as a result of such foresight, materially secure generates the leisure time necessary for social and technological innovation.

The other suspected function of the frontal lobes is to make possible mankind's bipedal posture. Our upright stance may not have been possible before the development of the frontal lobes. As we shall see later in more detail, standing on our own two feet freed our hands for manipulation, which then led to a major accretion of human cultural and physiological traits. In a very real sense, civilization may be a product of the frontal lobes.

Visual information from the eyes arrives in the human brain chiefly in the occipital lobe, in the back of the head; auditory impressions, in the upper part of the temporal lobe, beneath the temple. There is fragmentary evidence that these components of the neocortex are substantially less well developed in blind deaf-mutes. Lesions in the occipital lobe—as produced by gunshot wounds, for example—frequently induce an impairment in the field of vision. The victim may be in all other respects normal but able to see only with peripheral vision, perceiving a solid, dark blot looming in front of him at the center of the normal field of view. In other cases, more bizarre perceptions follow, including geometrically regular, cursive floating impairments in the visual field, and "visual fits" in which (for example) objects on the floor to the patient's lower right are momentarily perceived as floating in the air to his upper left and rotated 180 degrees through space. It may even be possible to map which parts of the occipital lobes are responsible for which visual functions by systematically calculating the impairments of vision from various occipital lesions. Permanent impairments of vision are much less likely to occur in the very young, whose brains seem able to repair themselves or transfer functions to neighboring regions very well.

The ability to connect auditory with visual stimuli is also localized in the temporal lobe. Lesions in the temporal lobe can result in a form of aphasia, the inability to recognize spoken words. It is remarkable and significant that brain-damaged patients can be completely competent in spoken language and entirely incompetent in written language, or vice versa. They may be able to write but unable to read; able to read numbers but not letters; able to name objects but not colors. There is in the neocortex a striking separation of function, which is contrary to such common-sense notions as that reading and writing, or recognizing words and numbers, are very similar activities. There are also as yet unconfirmed reports of brain damage that results only in the inability to understand the passive voice or prepositional phrases or possessive constructions. (Perhaps the locale of the subjunctive mood will one day be found. Will Latins turn out to be extravagantly endowed and English-speaking peoples significantly short-changed in this minor piece of brain anatomy?) Various abstractions, including the "parts of speech" in grammar, seem, astonishingly, to be wired into specific regions of the brain.

In one case, a temporal-lobe lesion resulted in a surprising impairment in the patient's perception of faces, even the faces of his immediate family. Presented with the face on the opposite page, he described it as "possibly" being an apple. Asked to justify this interpretation, he identified the mouth as a cut in the apple, the nose as the stem of the apple folded back on its surface, and the eyes as two worm holes. The same patient was perfectly able to recognize sketches of houses and other inanimate objects. A wide range of experiments

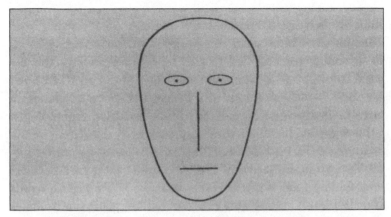

Face described by a patient as an apple. (Otherwise: apple described by a physician as a face.) After Teuber.

shows that lesions in the right temporal lobe produce amnesia for certain types of nonverbal material, while lesions in the left temporal lobe produce a characteristic loss of memory for language.

Our ability to read and make maps, to orient ourselves spatially in three dimensions and to use the appropriate symbols—all of which are probably involved in the development if not the use of language— are severely impaired by lesions in the parietal lobes, near the top of the head. One soldier who suffered a massive wartime penetration of the parietal lobe was for a full year unable to orient his feet into his slippers, much less find his bed in the hospital ward. He nevertheless eventually experienced an almost complete recovery.

A lesion of the angular gyrus of the neocortex, in the parietal lobe, results in alexia, the inability to recognize the printed word. The parietal lobe appears to be involved in all human symbolic language and, of all the brain lesions, a lesion in the parietal lobe causes the greatest decline in intelligence as measured by activities in everyday life.

Chief among the neocortical abstractions are the human symbolic languages, particularly reading and writing and mathematics.

These seem to require cooperative activities of the temporal, parietal and frontal lobes, and perhaps the occipital as well. Not all symbolic languages are neocortical however; bees—without a hint of a neocortex—have an elaborate dance language, first elucidated by the Austrian entomologist Karl von Frisch, by which they communicate information on the distance and direction of available food. It is an exaggerated gestural language, imitative of the motions bees in fact perform when finding food—as if we were to make a few steps towards the refrigerator, point and rub our bellies, with our tongues lolling out all the while. But the vocabularies of such languages are extremely limited, perhaps only a few dozen words. The kind of learning that human youngsters experience during their long childhood seems almost exclusively a neocortical function.

While most olfactory processing is in the limbic system, some occurs in the neocortex. The same division of function seems to apply to memory. A principal part of the limbic system, other than the olfactory cortex, is, as we have mentioned, the hippocampal cortex. When the olfactory cortex is excised, animals can still smell, although with a much lower efficiency. This is another demonstration of the redundancy of brain function. There is some evidence that, in contemporary humans, the short-term memory of smell resides in the hippocampus. The original function of the hippocampus may have been exclusively the short-term memory of smell, useful in, for example, tracking prey or finding the opposite sex. But a bilateral hippocampal lesion in humans results, as in the case of H. M., in a profound impairment of all varieties of short-term memory. Patients with such lesions literally cannot remember from one moment to the next. Clearly, both hippocampus and frontal lobes are involved in human short-term memory.

One of the many interesting implications of this is that short-term and long-term memory reside mostly in different parts of the brain. Classical conditioning—the ability of Pavlov's dogs to salivate when the bells rang—seems to be located in the limbic system. This is long-term memory, but of a very restricted kind. The sophisticated sort of human long-term memory is situated in the neocortex, which is consistent with the human ability to think ahead. As we grow old, we sometimes forget what has just been said to us while retaining vivid and accurate recollections of events in our childhood. In such cases, little seems to be wrong with either our short-term or our long-term memories; the problem is the connection between the two—we have great difficulty in accessing new material into the long-term memory. Penfield believed that this lost accessing ability arises from an inadequate blood supply to the hippocampus in old age—because of arteriosclerosis or other physical disabilities. Thus elderly people—and ones not so elderly—may have serious impairments in accessing short-term memory while being otherwise perfectly alert and intellectually keen.* This phenomenon also shows a clear-cut distinction between short-term and long-term memory, consistent with their localization in different parts of the brain. Waitresses in short-order restaurants can remember an impressive amount of information, which they accurately transmit to the kitchen. But an hour later, the information has vanished utterly. It was put into the short-term memory only, and no effort was made to access it into the long-term memory.

The mechanics of recall can be complex. A common experience is that we know something is in our long-term memory—a word, a name, a face, an experience—but find ourselves unable to call it up. No matter how hard we try, the memory resists retrieval. But if we think sideways at it, recalling some slightly related or peripheral item, it often follows unbidden. (Human vision is also a little like this. When we look directly at a faint object—a star, say—we are using the

*Indeed, there is a range of medical evidence on the connection between blood supply and intellectual abilities. It has long been known that patients deprived of oxygen for some minutes can experience permanent and serious mental impairment. Operations to remove material from clogged carotid arteries in an effort to prevent stroke yield unexpected benefits. According to one study, six weeks after such operations, the patients showed an average increase in IQ of eighteen points, a substantial improvement. And there has been some speculation that immersion in hyperbaric oxygen—that is, oxygen under high pressure—can raise the intelligence of infants.

fovea, the part of the retina with the greatest acuity and the greatest
density of cells called cones. But when we avert our vision slightly—
when, in a manner of speaking, we look sideways at the object—we
bring into play the cells called rods, which are sensitive to feeble illu-
mination and so able to perceive the faint star.) It would be interesting
to know why thinking sideways improves memory retrieval; it may
be merely associating to the memory trace by a different neural path-
way. But it does not suggest particularly efficient brain engineering.

We have all had the experience of awakening with a particularly vivid, chilling, insightful or otherwise memorable dream clearly
in mind; saying to ourselves, "I'll certainly remember *this* dream in
the morning"; and the next day having not the foggiest notion about
the content of the dream or, at best, a vague trace of an emotion
tone. On the other hand, if I am sufficiently exercised about the
dream to awaken my wife in the middle of the night and tell her
about it, I have no difficulty remembering its contents unaided in
the morning. Likewise, if I take the trouble of writing the dream
down, when I awaken the next morning I can remember the dream
perfectly well without referring to my notes. The same thing is true
of, for example, remembering a telephone number. If I am told a
number and merely think about it, I am likely to forget it or trans-
pose some of the digits. If I repeat the numbers out loud or write
them down, I can remember them quite well. This surely means
that there is a part of our brain which remembers sounds and images,
but not thoughts. I wonder if that sort of memory arose before we
had very many thoughts—when it was important to remember the
hiss of an attacking reptile or the shadow of a plummeting hawk,
but not our own occasional philosophical reflections.

ON HUMAN NATURE

Despite the intriguing localization of function in the triune-
brain model, it is, I stress again, an oversimplification to insist upon
perfect separation of function. Human ritual and emotional

behavior are certainly influenced strongly by neocortical abstract reasoning; analytical demonstrations of the validity of purely religious beliefs have been proffered, and there are philosophical justifications for hierarchical behavior, such as Thomas Hobbes' "demonstration" of the divine right of kings. Likewise, animals that are not human—and in fact even some animals that are not primates—seem to show glimmerings of analytical abilities. I certainly have such an impression about dolphins, as I described in my book *The Cosmic Connection.*

Nevertheless, while bearing these caveats in mind, it seems a useful first approximation to consider the ritualistic and hierarchical aspects of our lives to be influenced strongly by the R-complex and shared with our reptilian forebears; the altruistic, emotional and religious aspects of our lives to be localized to a significant extent in the limbic system and shared with our nonprimate mammalian forebears (and perhaps the birds); and reason to be a function of the neocortex, shared to some extent with the higher primates and such cetaceans as dolphins and whales. While ritual, emotion and reasoning are all significant aspects of human nature, the most nearly unique human characteristic is the ability to associate abstractly and to reason. Curiosity and the urge to solve problems are the emotional hallmarks of our species; and the most characteristically human activities are mathematics, science, technology, music and the arts— a somewhat broader range of subjects than is usually included under the "humanities." Indeed, in its common usage this very word seems to reflect a peculiar narrowness of vision about what is human. Mathematics is as much a "humanity" as poetry. Whales and elephants may be as "humane" as humans.

The triune-brain model derives from studies of comparative neuroanatomy and behavior. But honest introspection is not unknown in the human species, and if the triune-brain model is correct, we would expect some hint of it in the history of human self-knowledge. The most widely known hypothesis that is at least reminiscent of the triune brain is Sigmund Freud's division of the human

Mosaic II by M. C. Escher.

psyche into id, ego and superego. The aggressive and sexual aspects
of the R-complex correspond satisfyingly to the Freudian description of the id (Latin for "it"—i.e., the beast-like aspect of our natures);
but, so far as I know, Freud did not in his description of the id lay
great stress on the ritual or social-hierarchy aspects of the R-complex. He did describe emotions as an ego function—in particular
the "oceanic experience," which is the Freudian equivalent of the
religious epiphany. However, the superego is not depicted primarily as the site of abstract reasoning but rather as the internalizer of
societal and parental strictures, which in the triune brain we might
suspect to be more a function of the R-complex. Thus I would have
to describe the psychoanalytic tripartite mind as only weakly in
accord with the triune-brain model.

Perhaps a better metaphor is Freud's division of the mind into the conscious; the preconscious, which is latent but capable of being tapped; and the unconscious, which is repressed or otherwise unavailable. It was the tension that exists among the components of the psyche that Freud had in mind when he said of man that "his capacity for neurosis would merely be the obverse of his capacity for cultural development." He called the unconscious functions "primary processes."

A superior agreement is found in the metaphor for the human psyche in the Platonic dialogue *Phaedrus.* Socrates likens the human soul to a chariot drawn by two horses—one black, one white—pulling in different directions and weakly controlled by a charioteer. The metaphor of the chariot itself is remarkably similar to MacLean's neural chassis; the two horses, to the R-complex and the limbic cortex; and the charioteer barely in control of the careening chariot and horses, to the neocortex. In yet another metaphor, Freud described the ego as the rider of an unruly horse. Both the Freudian and the Platonic metaphors emphasize the considerable independence of and tension among the constituent parts of the psyche, a point that characterizes the human condition and to which we will return. Because of the neuroanatomical connections between the three components, the triune brain must itself, like the *Phaedrus* chariot, be a metaphor; but it may prove to be a metaphor of great utility and depth.

Then wilt thou not be loth
To leave this Paradise, but shalt possess
A Paradise within thee, happier far ...
They hand in hand with wandering steps and slow
Through Eden took their solitary way.

JOHN MILTON
Paradise Lost

Why didst thou leave the trodden paths of men
Too soon, and with weak hands though mighty heart
Dare the unpastured dragon in his den?
Defenseless as thou wert, oh, where was then
Wisdom, the mirrored shield ...?

PERCY BYSSHE SHELLEY
Adonais

CHAPTER IV

EDEN
AS A METAPHOR:
THE EVOLUTION
OF MAN

FOR THEIR surface area, insects weigh very little. A beetle, falling from a high altitude, quickly achieves terminal velocity: air resistance prevents it from falling very fast, and, after alighting on the ground, it will walk away, apparently none the worse for the experience. The same is true of small mammals—squirrels, say. A mouse can be dropped down a thousand-foot mine shaft and, if the ground is soft, will arrive dazed but essentially unhurt. In contrast, human beings are characteristically maimed or killed by any fall of more than a few dozen feet: because of our size, we weigh too much for our surface area. Therefore our arboreal ancestors had to pay attention. Any error in brachiating from branch to branch could be fatal. Every leap was an opportunity for evolution. Powerful selective forces were at work to evolve organisms with grace and agility, accurate binocular vision, versatile manipulative abilities, superb eye–hand coordination, and an intuitive grasp of Newtonian gravitation. But each of these skills required significant advances in the evolution of the brains and particularly the neocortices of our ancestors. Human intelligence is fundamentally indebted to the millions of years our ancestors spent aloft in the trees.

And after we returned to the savannahs and abandoned the trees, did we long for those great graceful leaps and ecstatic moments of weightlessness in the shafts of sunlight of the forest roof? Is the startle reflex of human infants today to prevent falling from the treetops? Are our nighttime dreams of flying and our daytime passion for flight, as exemplified in the lives of Leonardo da Vinci or Konstantin Tsiolkovskii, nostalgic reminiscences of those days gone by in the branches of the high forest?*

Other mammals, even other nonprimate and noncetacean mammals, have neocortices. But in the evolutionary line leading to man, when was the first large-scale development of the neocortex? While none of our simian ancestors are still around, this question can nevertheless be answered or at least approached: we can examine fossil skulls. In humans, in apes and monkeys, and in other mammals, the brain volume almost fills the skull. This is not true, for example, in fish. Thus by taking a cast of a skull, we can determine what

---

*Modern rocket technology and space exploration owes an incalculable debt to Dr. Robert H. Goddard, who through many decades of devoted and lonely research was singlehandedly responsible for the development of essentially all important aspects of the modern rocket. Goddard's interest in this subject originated in a magic moment. In the New England autumn of 1899, Goddard was a seventeen-year-old high school sophomore who had climbed a cherry tree and, while idly looking down at the ground around him, experienced a kind of epiphanal vision of a vehicle that would transport human beings to the planet Mars. He resolved to devote himself to the task. Exactly one year later, he climbed the tree again, and on every October 19th for the rest of his life, made a special point of recollecting that moment. Can it be an accident that this vision of voyages to the planets, which has led directly to its own historical fulfillment, was glimpsed in the limbs of a tree?

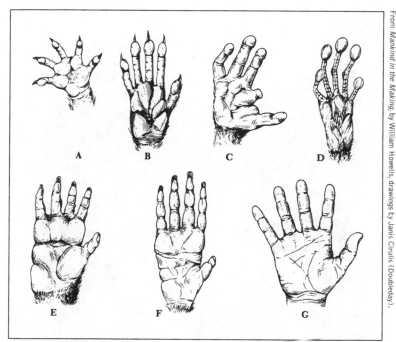

The hands of animals are adapted to their life styles, and vice versa. Shown are A the opossum; B the tree shrew; C the potto; D the tarsier; E the baboon (where this appendage is used partly as a hand and partly as a foot); F the orangutan, specialized for brachiation; and G humans, with a relatively long and opposable thumb.

is called the endocranial volume of our immediate ancestors and collateral relatives and can make some rough estimates of their brain volumes.

The question of who was and who was not an ancestor of man is still being hotly debated by the paleontologists, and hardly a year goes by without the discovery of some fossil of remarkably human aspect much older than anyone had previously thought possible. What seems certain is that about five million years ago, there was an abundance of apelike animals, the gracile Australopithecines, who

A family of gracile Australopithecines five million years ago.

walked on two feet and had brain volumes of about 500 cubic centimeters, some 100 cc more than the brain of a modern chimpanzee. With this evidence, paleontologists have deduced that "bipedalism preceded encephalization," by which they mean that our ancestors walked on two legs before they evolved big brains.

By three million years ago, there was a variety of bipedal fellows with a wide range of cranial volumes, some considerably larger than the East African gracile Australopithecines of a few million years earlier. One of them, which L. S. B. Leakey, the Anglo-Kenyan student of early man, called *Homo habilis*, had a brain volume of about 700 cubic centimeters. We also have archaeological evidence that *Homo habilis* made tools. The idea that tools are both the cause and the effect of walking on two legs, which frees the hands, was first advanced by Charles Darwin. The fact that these significant changes in behavior are accompanied by equally significant changes in brain volume does not prove that the one is caused by the other; but our previous discussion makes such a causal link appear very likely.

The table on page 80 summarizes the fossil evidence, through 1976, on our most recent ancestors and collateral relatives. The two rather different kinds of Australopithecines were not of the genus *Homo*, not human; they were still incompletely bipedal and had brain masses only about a third the size of the average adult human brain today. Were we to meet an Australopithecine, say, on the subway, we would perhaps be struck most by the almost total absence of forehead. He was the lowest of lowbrows. There are significant differences between the two kinds of Australopithecines. The robust species was taller and heavier, with most impressive "nutcracker" teeth and a remarkable evolutionary stability. The endocranial volume of *A. robustus* varies very little from specimen to specimen over millions of years of time. The gracile Australopithecines, judging again from their teeth, probably ate meat as well as vegetables. They were smaller and lither, as their name indicates. However, they are considerably older and have much more variance in endocranial volume than their robust cousins. But, most important, the gracile

RECENT ANCESTORS AND COLLATERAL RELATIVES

Species	Earliest Specimen	Endocranial Volume	Height and Weight	Ratio, Body to Brain Weight	Comments
Australopithecus robustus (including Paranthropus and Zinjanthropus)	3.5 m.y.	500-550 cc	1.5 m (5') 40-60 kg (85-130 lbs.)	~ 90	Powerful masticatory apparatus; sagittal crest; probably rigid vegetarian; imperfectly biped; no forehead. Bush habitat. No associated tools.
Australopithecus africanus (gracile Australopithecine)	6 m.y.	430-600 cc	1-1.25 m (3'-4') 20-30 kg (45-65 lbs.)	~ 50	Stronger canines and incisors; probable omnivores; imperfectly biped; slight forehead. Bush and brush habitat. Stone and bone tools.
Homo habilis	3.7 m.y.	500-800 cc	1.2-1.4 m (4'-4½') 30-50 kg (65-110 lbs.)	~ 60	High forehead. Definite omnivore. Completely bipedal. Savannah habitat. Stone tools, possible building construction.
Homo erectus (Pithecanthropus)	1.5 m.y.	750-1250 cc	1.4-1.8 m (4½'-6') 40-80 kg (100-180 lbs.)	~ 65	High forehead. Definite omnivore. Completely bipedal. Varied habitat. Varied stone tools. Invention of fire.
Homo sapiens	0.2 m.y.	1100-2200 cc	1.4-2 m (4½'-6½') 40-100 kg (100-220 lbs.)	~ 45	High forehead. Definite omnivore. Completely bipedal. Global habitat. Stone, metal, chemical, electronic, nuclear tools.

m.y. = million years; cc = cubic centimeters; m = meters; kg = kilograms

Australopithecine sites are associated with a clear industry: the manufacture of tools made of stone and animal bones, horns and teeth—painstakingly carved, broken, rubbed and polished to make chipping, flaking, pounding and cutting tools. No tools have been associated with *A. robustus.* The ratio of brain weight to body weight is almost twice as large for the gracile as for the robust Australopithecus, and it is a natural speculation to wonder whether that factor of two is the difference between tools and no tools.

At apparently the same epoch as the emergence of *Australopithecus robustus,* there arose a new animal, *Homo habilis,* the first true man. He was larger, both in body and in brain weight, than either of the Australopithecines, and had a ratio of brain to body weight about the same as that of the gracile Australopithecines. He emerged at a time when, for climatic reasons, the forests were receding. *Homo habilis* inhabited the vast African savannahs, an extremely challenging environment filled with an enormous variety of predators and prey. On these plains of low grass appeared both the first modern man and the first modern horse. They were almost exact contemporaries.

In the last sixty million years, there has been a continuous evolution of ungulates, well recorded in the fossil record, and eventually culminating in the modern horse. Eohippus, the "dawn horse" of some fifty million years ago, was about the size of an English collie, with a brain volume of about twenty-five cubic centimeters, and a ratio of brain to body weight about half that of comparable contemporary mammals. Since then, horses have experienced a dramatic evolution in both absolute and relative brain size, with major developments in the neocortex and particularly in the frontal lobes—an evolution certainly accompanied by major improvements in equine intelligence. I wonder if the parallel developments in the intelligence of horse and man might have a common cause. Did horses, for example, have to be swift of foot, acute of sense, and intelligent to elude predators which hunted primate as well as equine prey?

The East African savannah near Olduvai Gorge a few million years ago. In right foreground are three hominids, perhaps Australopithecines, perhaps *Homo habilis*. The active volcano in the background is now Mt. Ngorongoro.

H. habilis had a high forehead, suggesting a significant development of the neocortical areas in the frontal and temporal lobes as well as the regions in the brain, to be discussed later, that seem to be connected with the power of speech. Were we to encounter *Homo habilis*—dressed, let us say, in the latest fashion on the boulevards of some modern metropolis—we would probably give him only a passing glance, and that because of his relatively small stature. Associated with *Homo habilis* are a variety of tools of considerable sophistication. In addition, there is evidence from various circular arrangements of stones that *Homo habilis* may have constructed dwellings; that long before the Pleistocene Ice Ages, long before men regularly inhabited caves, *H. habilis* was constructing homes out-of-doors—probably of wood, wattle, grass and stone.

Since *H. habilis* and *A. robustus* emerged at the same time, it is very unlikely that one was the ancestor of the other. The gracile

Australopithecines were also contemporaries of *Homo habilis* but much more ancient. It is therefore possible—although by no means certain—that both *H. habilis*, with a promising evolutionary future, and *A. robustus*, an evolutionary dead end, arose from the gracile *A. africanus*, who survived long enough to be their contemporary.

The first man whose endocranial volume overlaps that of modern humans is *Homo erectus*. For many years the principal specimens of *H. erectus* were known from China and thought to be about half a million years old. But in 1976 Richard Leakey of the National Museums of Kenya reported a nearly complete skull of *Homo erectus* found in geological strata one and a half million years old. Since the Chinese specimens of *Homo erectus* are clearly associated with the remains of campfires, it is possible that our ancestors domesticated fire much more than one half million years ago—which makes Prometheus far older than many had thought.

Perhaps the most striking aspect of the archaeological record concerning tools is that as soon as they appear at all they appear in enormous abundance. It looks very much as though an inspired gracile Australopithecine discovered for the first time the use of tools and immediately taught the toolmaking skill to his relatives and friends. There is no way to explain the discontinuous appearance of stone tools unless the Australopithecines had educational institutions. There must have been some sort of stonecraft guild passing on from generation to generation the precious knowledge about the fabrication and use of tools—knowledge that would eventually propel such feeble and almost defenseless primates into domination of the planet Earth. Whether the genus *Homo* independently invented tools or borrowed the discovery from the genus *Australopithecus* is not known.

We see from the table that the ratio of body to brain weight is, within the variance of measurement, roughly the same for the gracile Australopithecines, *Homo habilis*, *Homo erectus* and modern humans. The advances we have made in the last few million years cannot therefore be explained by the ratio of brain to body mass, but rather by increasing total brain mass, improved specialization of new function and complexity within the brain, and—especially—extrasomatic learning.

L. S. B. Leakey emphasized that the fossil record of a few million years ago is replete with a great variety of manlike forms, an interesting number of which are found with holes or fractures in their skulls. Some of these injuries may have been inflicted by leopards or hyenas; but Leakey and the South African anatomist Raymond Dart believed that many of them were inflicted by our ancestors. In Pliocene/Pleistocene times there was almost certainly a vigorous competition among many manlike forms, of which only one line survived—the tool experts, the line that led to us. What role killing played in that competition remains an open question. The gracile Australopithecines were erect, agile, fleet and three and a half feet tall: "little people." I sometimes wonder whether our myths about gnomes, trolls, giants and dwarfs could possibly be a genetic or cultural memory of those times.

At the same time that the hominid cranial volume was undergoing its spectacular increase, there was another striking change in human anatomy; as the British anatomist Sir Wilfred Le Gros Clark of Oxford University has observed, there was a wholesale reshaping of the human pelvis. This was very likely an adaptation to permit the live birth of the latest model large-brained babies. Today, it is unlikely that any further substantial enlargement of the pelvic girdle in the region of the birth canal is possible without severely impairing the ability of women to walk efficiently. (At birth, girls already have a significantly larger pelvis and skeletal pelvic opening than do boys; another large increment in the size of the female pelvis occurs at puberty.) The parallel emergence of these two evolutionary events illustrates nicely how natural selection works. Those mothers with hereditary large pelvises were able to bear large-brained babies who because of their superior intelligence were able to compete successfully in adulthood with the smaller-brained offspring of mothers with smaller pelvises. He who had a stone axe was more likely to win a vigorous difference of opinion in Pleistocene times. More important, he was a more successful hunter. But the invention and continued manufacture of stone axes required larger brain volumes.

So far as I know, childbirth is generally painful in only one of the millions of species on Earth: human beings. This must be a consequence of the recent and continuing increase in cranial volume. Modern men and women have braincases twice the volume of *Homo habilis*. Childbirth is painful because the evolution of the human skull has been spectacularly fast and recent. The American anatomist C. Judson Herrick described the development of the neocortex in the following terms: "Its explosive growth late in phylogeny is one of the most dramatic cases of evolutionary transformation known to comparative anatomy." The incomplete closure of the skull at birth, the fontanelle, is very likely an imperfect accommodation to this recent brain evolution.

The connection between the evolution of intelligence and the pain of childbirth seems unexpectedly to be made in the Book of Genesis. In punishment for eating the fruit of the tree of the knowledge of good and evil, God says to Eve,* "In pain shalt thou bring forth children" (Genesis 3:16). It is interesting that it is not the getting of *any* sort of knowledge that God has forbidden, but, specifically, the knowledge of the difference between good and evil—that is, abstract and moral judgments, which, if they reside anywhere, reside in the neocortex. Even at the time that the Eden story was written, the development of cognitive skills was seen as endowing man with godlike powers and awesome responsibilities. God says: "Behold, the

---

*God's judgment on the serpent is that henceforth "upon thy belly shalt thou go"— implying that previously reptiles traveled by an alternative mode of locomotion. This is, of course, precisely true: snakes have evolved from four-legged reptilian ancestors resembling dragons. Many snakes still retain anatomical vestiges of the limbs of their ancestors.

man is become as one of us, to know good and evil; and now, lest
he put forth his hand, and take also of the Tree of Life, and eat, and
live forever" (Genesis 3:22), he must be driven out of the Garden.
God places cherubim with a flaming sword east of Eden to guard
the Tree of Life from the ambitions of man.*

Perhaps the Garden of Eden is not so different from Earth as it
appeared to our ancestors of some three or four million years ago,
during a legendary golden age when the genus *Homo* was perfectly interwoven with the other beasts and vegetables. After the exile
from Eden we find, in the biblical account, mankind condemned
to death; hard work; clothing and modesty as preventatives of sexual stimulation; the dominance of men over women; the domestication
of plants (Cain); the domestication of animals (Abel); and murder
(Cain plus Abel). These all correspond reasonably well to the historical and archaeological evidence. In the Eden metaphor, there is
no evidence of murder before the Fall. But those fractured skulls of
bipeds not on the evolutionary line to man may be evidence that
our ancestors killed, even in Eden, many manlike animals.

Civilization develops not from Abel, but from Cain the mur-
derer. The very word "civilization" derives from the Latin word for
city. It is the leisure time, community organization and specialization of labor in the first cities that permitted the emergence of
the arts and technologies we think of as the hallmarks of civilizations. The first city, according to Genesis, was constructed by Cain,
the inventor of agriculture—a technology that requires a fixed
abode. And it is his descendants, the sons of Lamech, who invent
both "artifices in brass and iron" and musical instruments. Metal-
lurgy and music—technology and art—are in the line from Cain.
And the passions that lead to murder do not abate: Lamech says,
"For I have slain a man for wounding me, and a young man for

---

*Cherubim is plural; Genesis 3:24 specifies one flaming sword. Presumably flaming
swords were in short supply.

The creation of Adam: A relief on the doors of the Church of St. Peter in Bologna by Jacopo della Quercia.

bruising me; if Cain shall be avenged sevenfold, truly Lamech seventy and sevenfold." The connection between murder and invention has been with us ever since. Both derive from agriculture and civilization.

The temptation of Eve and Adam by a reptile with a remarkably human head: A relief on the doors of St. Peter in Bologna by Jacopo della Quercia.

One of the earliest consequences of the anticipatory skills that accompanied the evolution of the prefrontal lobes must have been the awareness of death. Man is probably the only organism on Earth with a relatively clear view of the inevitability of his own end.

The expulsion from Eden: A relief on the doors of St. Peter in Bologna by Jacopo della Quercia.

Burial ceremonies that include the interment of food and artifacts along with the deceased go back at least to the times of our Neanderthal cousins, suggesting not only a widespread awareness of death but also an already developed ritual ceremony to sustain the deceased in the afterlife. It is not that death was absent before the spectacular growth of the neocortex, before the exile from Eden; it is only that, until then, no one had ever noticed that death would be his destiny.

The fall from Eden seems to be an appropriate metaphor for some of the major biological events in recent human evolution. This may account for its popularity.* It is not so remarkable as to require us to believe in a kind of biological memory of ancient historical events, but it does seem to me close enough to risk at least raising the question. The only repository of such a biological memory is, of course, the genetic code.

By fifty-five million years ago, in the Eocene Period, there was a great proliferation of primates, both arboreal and ground-dwelling, and the evolution of a line of descent that eventually led to Man. Some primates of those times—e.g., a prosimian called *Tetonius*— exhibit in their endocranial casts tiny nubs where the frontal lobes will later evolve. The first fossil evidence of a brain of even vaguely human aspect dates back to eighteen million years to the Miocene Period, when an anthropoid ape which we call *Proconsul* or *Dryopithecus* appeared. *Proconsul* was quadrupedal and arboreal, probably ancestral to the present great apes and possibly to *Homo sapiens* as well. He is roughly what we might expect for a common ancestor of apes and men. (His approximate contemporary, *Ramapithecus*, is thought by some anthropologists to be ancestral to man.) *Proconsul*'s endocranial casts show recognizable frontal lobes but much less well developed neocortical convolutions than are displayed by apes and men today. His cranial volume was still very small. The biggest burst of evolution in cranial volume occurred in the last few million years.

---

*In the West. There are, of course, many insightful and profound myths on the origin of mankind in other human cultures.

Patients who have had prefrontal lobotomies have been described as losing a "continuing sense of self"—the feeling that I am a particular individual with some control over my life and circumstances, the "me-ness" of me, the uniqueness of the individual. It is possible that lower mammals and reptiles, lacking extensive frontal lobes, also lack this sense, real or illusory, of individuality and free will, which is so characteristically human and which may first have been experienced dimly by *Proconsul.*

The development of human culture and the evolution of those physiological traits we consider characteristically human most likely proceeded—almost literally—hand in hand: the better our genetic predispositions for running, communicating and manipulating, the more likely we were to develop effective tools and hunting strategies; the more adaptive our tools and hunting strategies, the more likely it was that our characteristic genetic endowments would survive. The American anthropologist Sherwood Washburn of the University of California, a principal exponent of this view, has said: "Much of what we think of as human evolved long after the use of tools. It is probably more correct to think of much of our structure as the result of culture than it is to think of men anatomically like ourselves slowly developing culture."

Some students of human evolution believe that part of the selection pressure behind this enormous burst in brain evolution was in the motor cortex and not at first in the neocortical regions responsible for cognitive processes. They stress the remarkable abilities of human beings to throw projectiles accurately, to move gracefully, and—as Louis Leakey enjoyed illustrating by direct demonstration—naked, to outrun and immobilize game animals. Such sports as baseball, football, wrestling, track and field events, chess and warfare may owe their appeal—as well as their largely male following—to these prewired hunting skills, which served us so well for millions of years of human history but which find diminished practical applications today.

Effective defense against predators and the hunting of game were both necessarily cooperative ventures. The environment in

which man evolved—in Africa in Plicene and Pleistocene times—
was inhabited by a variety of terrifying mammalian carnivores,
perhaps the most awesome of which were packs of large hyenas. It
was very difficult to defend oneself alone against such a pack. Stalking large animals, either solitary beasts or herds, is dangerous; some
gestural communication among the hunters is necessary. We know,
for example, that shortly after man entered North America, via the
Bering Straits in the Pleistocene Period, there were massive and spectacular kills of large game animals, often by driving them over cliffs.
In order to stalk a single wildebeest or stampede a herd of antelope
to their deaths, hunters must share at least a minimal symbolic language. Adam's first act was linguistic—long before the Fall and even
before the creation of Eve: he named the animals of Eden.

Some forms of gestural symbolic language, of course, originated much earlier than the primates; canines and many other mammals
who form dominance hierarchies may indicate submission by avert-
ing the eyes or baring the neck. We have mentioned other submissive
rituals in primates such as macaques. The human greetings of bow,
nod and curtsy may have a similar origin. Many animals seem to
signal friendship by biting, but not hard enough to hurt, as if to
say, "I am able to bite you but choose not to do so." The raising of
the right hand as a symbol of greeting among humans has precisely
the same significance: "I could attack you with a weapon but choose
not to wield one."*

---

*The upraised and open right hand is sometimes described as a "universal" symbol of
good will. It at least runs the gamut from Praetorian Guards to Sioux scouts. Since those
wielding weapons are, in human history, characteristically male, it should be and is a
characteristically male greeting. For these reasons, among others, the plaque aboard the
Pioneer 10 spacecraft—the first artifact of mankind to leave the solar system—included a drawing of a naked man and woman, the man's hand raised, palm out, in greeting
(see illustration on p. 221). In *The Cosmic Connection* I describe the humans on the
plaque as the most obscure part of the message. Nevertheless, I wonder. Could the significance of the man's gesture be deduced by beings with very different biologies?

Photo by Nat Farbman, *Life.* Courtesy of Time-Life Picture Agency, © Time Inc.

The development of human language was a crucial turning point in the evolution of man. Among its highest peaks, as here, were story-telling cultures before the invention of writing.

Extensive gestural languages were employed by many human hunting communities—for example, among the Plains Indians, who also used smoke signals. According to Homer, the victory of the Hellenes at Troy was conveyed from Ilium to Greece, a distance of some hundreds of miles, by a series of signal fires. The date was about 1100 B.C. However, both the repertoire of ideas and the speed with which ideas can be communicated in gestural or sign languages is limited. Darwin pointed out that gestural languages cannot usefully be employed while our hands are otherwise occupied, or at night, or when our view of the hands is obstructed. One can imagine gestural languages being gradually supplemented and then supplanted by verbal languages—which originally may have been onomatopoeic (that is, imitative in sound of the object or action being described). Children call dogs "bow-wows." In almost all human languages the child's word for "mother" seems imitative of the sound made inadvertently while feeding at the breast. But all of this could not have occurred without a restructuring of the brain.

We know from skeletal remains associated with early man that our ancestors were hunters. We know enough about the hunting of large animals to realize that some language is required for cooperative stalking. But ideas on the antiquity of language have received a measure of unexpected support from detailed studies of fossil endocasts made by the American anthropologist Ralph L. Holloway of Columbia University. Holloway's casts of fossil skulls are made of rubber latex, and he has attempted to deduce something of the detailed morphology of the brain from the shape of the skull. The activity is a kind of phrenology, but on the inside rather than on the outside and much more soundly based. Holloway believes that a region of the brain known as Broca's area, one of several centers required for speech, can be detected in fossil endocasts; and that he has found evidence for Broca's area in a *Homo habilis* fossil more than two million years old. The development of language, tools and culture may have occurred roughly simultaneously.

There were, incidentally, manlike creatures who lived only a few tens of thousands of years ago—the Neanderthals and the Cro-Magnons—who had average brain volumes of about 1,500 cubic centimeters; that is, more than a hundred cubic centimeters larger than ours. Most anthropologists guess that we are not descended from Neanderthals and may not be from Cro-Magnons either. But their existence raises the question: Who were those fellows? What were their accomplishments? Cro-Magnon was also very large: some specimens were well over six feet tall. We have seen that a difference in brain volume of 100 cubic centimeters does not seem to be significant, and perhaps they were no smarter than we or our immediate

A Pleistocene summit. Left to right: *Homo habilis* (in an inadequate state of repair), *Homo erectus*, Neanderthal man, Cro-Magnon man, and *Homo sapiens*.

ancestors; or perhaps they had other, still unknown, physical impediments. Neanderthal was a lowbrow, but his head was long, front to back; in contrast, our heads are not so deep, but they are taller: we can certainly be described as highbrows. Might the brain growth exhibited by Neanderthal man have been in the parietal and occipital lobes, and the major brain growth of our ancestors in the frontal and temporal lobes? Is it possible that the Neanderthals developed quite a different mentality than ours, and that our superior linguistic and anticipatory skills enabled us to destroy utterly our husky and intelligent cousins?

So far as we know, nothing like human intelligence appeared on Earth before a few million, or at least a few tens of millions of years ago. But that is a few tenths of a percent of the age of Earth, very late in December in the Cosmic Calendar. Why did it appear so late? The answer clearly seems to be that some particular property of higher primate and cetacean brains did not evolve until recently. But what is that property? I can suggest at least four possibilities, all of which have already been mentioned, either explicitly or implicitly: (1) Never before was there a brain so massive; (2) Never before was there a brain with so large a ratio of brain to body mass; (3) Never before was there a brain with certain functional units (large frontal and temporal lobes, for example); (4) Never before was there a brain with so many neural connections or synapses. (There seems to be some evidence that along with the evolution of the human brain there may have been an increase in the number of connections of each neuron with its neighbor, and in the number of microcircuits.) Explanations 1, 2 and 4 argue that a quantitative change produced a qualitative change. It does not seem to me that a crisp choice among these four alternatives can be made at the present time, and I suspect that the truth will actually embrace most or all of these possibilities.

The British student of human evolution Sir Arthur Keith proposed what he called a "Rubicon" in the evolution of the human brain. He thought that at the brain volume of *Homo erectus*—about 750 cubic centimeters, roughly the engine displacement of a fast motorcycle—the uniquely human qualities begin to emerge. The "Rubicon" might, of course, have been more qualitative than quantitative. Perhaps the difference was not so much an additional 200 cubic centimeters as some specific developments in the frontal, temporal and parietal lobes which provided us with analytical ability, foresight and anxiety.

While we can debate what the "Rubicon" corresponds to, the idea of some sort of Rubicon is not without value. But if there is a Rubicon anywhere near 750 cubic centimeters, while differences of the order of 100 or 200 cubic centimeters do not—at any rate to us—seem to be compelling determinants of intelligence, might not the apes be intelligent in some recognizably human sense? A typical chimpanzee brain volume is 400 cubic centimeters; a lowland gorilla's, 500 cc. This is the range of brain volumes among the tool-using gracile Australopithecines.

The Jewish historian Josephus added to the list of penalties and tribulations that accompanied Mankind's exile from Eden the loss of our ability to communicate with the animals. Chimpanzees have large brains; they have well-developed neocortices; they, too, have long childhoods and extended periods of plasticity. Are they capable of abstract thought? If they're smart, why don't they talk?

I demand of you, and of the whole world, that you show me a generic character ... by which to distinguish between Man and Ape. I myself most assuredly know of none. I wish somebody would indicate one to me. But, if I had called man an ape, or vice versa, I would have fallen under the ban of all the ecclesiastics. It may be that as a naturalist I ought to have done so.

CARL LINNAEUS,
the founder of taxonomy, *1788*

CHAPTER V

THE ABSTRACTIONS
OF BEASTS

Beast abstract not," announced John Locke, expressing mankind's prevailing opinion throughout recorded history. Bishop Berkeley had, however, a sardonic rejoinder: "If the fact that brutes abstract not be made the distinguishing property of that sort of animal, I fear a great many of those that pass for men must be reckoned into their number." Abstract thought, at least in its more subtle varieties, is not an invariable accompaniment of everyday life for the average man. Could abstract thought be a matter not of kind but of degree? Could other animals be capable of abstract thought but more rarely or less deeply than humans?

We have the impression that other animals are not very intelligent. But have we examined the possibility of animal intelligence carefully enough, or, as in François Truffaut's poignant film *The Wild Child*, do we simply equate the absence of our style of expression of intelligence with the absence of intelligence? In discussing communication with the animals, the French philosopher Montaigne remarked, "The defect that hinders communication betwixt them and us, why may it not be on our part as well as theirs?"*

There is, of course, a considerable body of anecdotal information suggesting chimpanzee intelligence. The first serious study of the behavior of simians—including their behavior in the wild—was made in Indonesia by Alfred Russel Wallace, the co-discoverer of evolution by natural selection. Wallace concluded that a baby orangutan he studied behaved "exactly like a human child in similar circumstances." In fact, "orangutan" is a Malay phrase meaning not ape but "man of the woods." Teuber recounted many stories told by his parents, pioneer German ethologists who founded and operated the first research station devoted to chimpanzee behavior on Tenerife in the Canary Islands early in the second decade of this century. It was here that Wolfgang Kohler performed his famous studies of Sultan, a chimpanzee "genius" who was able to connect two rods in order to reach an otherwise inaccessible banana. On Tenerife, also, two chimpanzees were observed maltreating a chicken: One would extend some food to the fowl, encouraging it to

---

*Our difficulties in understanding or effectuating communication with other animals may arise from our reluctance to grasp unfamiliar ways of dealing with the world. For example, dolphins and whales, who sense their surrounding with a quite elaborate sonar echo location technique, also communicate with each other by a rich and elaborate set of clicks, whose interpretation has so far eluded human attempts to understand it. One very clever recent suggestion, which is now being investigated, is that dolphin/dolphin communication involves a re-creation of the sonar reflection characteristics of the objects being described. In this view a dolphin does not "say" a single word for shark, but rather transmits a set of clicks corresponding to the audio reflection spectrum it would obtain on irradiating a shark with sound waves in the dolphin's sonar mode. The basic form of dolphin/dolphin communication in this view would be a sort of aural onomatopoeia, a drawing of audio frequency pictures—in this case, caricatures of a shark. We could well imagine the extension of such a language from concrete to abstract ideas, and by the use of a kind of audio rebus—both analogous to the development in Mesopotamia and Egypt of human written languages. It would also be possible, then, for dolphins to create extraordinary audio images out of their imaginations rather than their experience.

approach; whereupon the other would thrust at it with a piece of wire it had concealed behind its back. The chicken would retreat but soon allow itself to approach once again—and be beaten once again. Here is a fine combination of behavior sometimes thought to be uniquely human: cooperation, planning a future course of action, deception and cruelty. It also reveals that chickens have a very low capacity for avoidance learning.

Until a few years ago, the most extensive attempt to communicate with chimpanzees went something like this: A newborn chimp was taken into a household with a newborn baby, and both would be raised together—twin cribs, twin bassinets, twin high chairs, twin potties, twin diaper pails, twin baby-powder cans. At the end of three years, the young chimp had, of course, far outstripped the young human in manual dexterity, running, leaping, climbing and other motor skills. But while the child was happily babbling away, the chimp could say only, and with enormous difficulty, "Mama," "Papa," and "cup." From this it was widely concluded that in language, reasoning and other higher mental functions, chimpanzees were only minimally competent: "Beasts abstract not."

But in thinking over these experiments, two psychologists, Beatrice and Robert Gardner, at the University of Nevada realized that the pharynx and larynx of the chimp are not suited for human speech. Human beings exhibit a curious multiple use of the mouth for eating, breathing and communicating. In insects such as crickets, which call to one another by rubbing their legs, these three functions are performed by completely separate organ systems. Human spoken language seems to be adventitious. The exploitation of organ systems with other functions for communication in humans is also indicative of the comparatively recent evolution of our linguistic abilities. It might be, the Gardners reasoned, that chimpanzees have substantial language abilities which could not be expressed because of the limitations of their anatomy. Was there any symbolic language, they asked, that could employ the strengths rather than the weaknesses of chimpanzee anatomy?

The Gardners hit upon a brilliant idea: Teach a chimpanzee American sign language, known by its acronym Ameslan, and sometimes as "American deaf and dumb language" (the "dumb" refers, of course, to the inability to speak and not to any failure of intelligence). It is ideally suited to the immense manual dexterity of the chimpanzee. It also may have all the crucial design features of verbal languages.

There is by now a vast library of described and filmed conversations, employing Ameslan and other gestural languages, with Washoe, Lucy, Lana and other chimpanzees studied by the Gardners and others. Not only are there chimpanzees with working vocabularies of 100 to 200 words; they are also able to distinguish among nontrivially different grammatical patterns and syntaxes.

Washoe (left) signaling in Ameslan "hat," for a woolen cap.

What is more, they have been remarkably inventive in the construction of new words and phrases.

On seeing for the first time a duck land quacking in a pond, Washoe gestured "water bird," which is the same phrase used in English and other languages, but which Washoe invented for the occasion. Having never seen a spherical fruit other than an apple, but knowing the signs for the principal colors, Lana, upon spying a technician eating an orange, signed "orange apple." After tasting a watermelon, Lucy described it as "candy drink" or "drink fruit," which is essentially the same word form as the English "water melon." But after she had burned her mouth on her first radish, Lucy forever after described them as "cry hurt food." A small doll placed unexpectedly in Washoe's cup elicited the response "Baby in my drink." When Washoe soiled, particularly clothing or furniture, she

Washoe (left) signaling in Ameslan "sweet," for a lollipop.

was taught the sign "dirty," which she then extrapolated as a general term of abuse. A rhesus monkey that evoked her displeasure was repeatedly signed at: "Dirty monkey, dirty monkey, dirty monkey." Occasionally Washoe would say things like "Dirty Jack, gimme drink." Lana, in a moment of creative annoyance, called her trainer "You green shit." Chimpanzees have invented swear words. Washoe also seems to have a sort of sense of humor; once, when riding on her trainer's shoulders and, perhaps inadvertently, wetting him, she signed: "Funny, funny."

Lucy was eventually able to distinguish clearly the meanings of the phrases "Roger tickle Lucy" and "Lucy tickle Roger," both of which activities she enjoyed with gusto. Likewise, Lana extrapolated from "Tim groom Lana" to "Lana groom Tim." Washoe was observed "reading" a magazine—i.e., slowly turning the pages, peering intently at the pictures and making, to no one in particular, an appropriate sign, such as "cat" when viewing a photograph of a tiger, and "drink" when examining a Vermouth advertisement. Having learned the sign "open" with a door, Washoe extended the concept to a briefcase. She also attempted to converse in Ameslan with the laboratory cat, who turned out to be the only illiterate in the facility. Having acquired this marvelous method of communication, Washoe may have been surprised that the cat was not also competent in Ameslan. And when one day Jane, Lucy's foster mother, left the laboratory, Lucy gazed after her and signed: "Cry me. Me cry."

Boyce Rensberger is a sensitive and gifted reporter for the *New York Times* whose parents could neither speak nor hear, although he is in both respects normal. His first language, however, was Ameslan. He had been abroad on a European assignment for the *Times* for some years. On his return to the United States, one of his first domestic duties was to look into the Gardners' experiments with Washoe. After some little time with the chimpanzee, Rensberger reported, "Suddenly I realized I was conversing with a member of another species in my native tongue." The use of the word tongue is, of course, figurative: it is built deeply into the structure of the language (a word that also means "tongue"). In fact, Rensberger was conversing with a member of another species in his native "hand." And it is just this transition from tongue to hand that has permitted humans to regain the ability—lost, according to Josephus, since Eden—to communicate with the animals.

In addition to Ameslan, chimpanzees and other nonhuman primates are being taught a variety of other gestural languages. At the Yerkes Regional Primate Research Center in Atlanta, Georgia, they are learning a specific computer language called (by the humans, not the chimps) "Yerkish." The computer records all of its subjects' conversations, even during the night when no humans are in attendance; and from its ministrations we have learned that chimpanzees prefer jazz to rock and movies about chimpanzees to movies about human beings. Lana had, by January 1976, viewed *The Developmental Anatomy of the Chimpanzee* 245 times. She would undoubtedly appreciate a larger film library.

In the illustration on page 106, Lana is shown requesting, in proper Yerkish, a piece of banana from the computer. The syntax required to request from the computer water, juice, chocolate candy, music, movies, an open window and companionship are also displayed. (The machine provides for many of Lana's needs, but not all. Sometimes, in the middle of the night, she forlornly types out: "Please, machine, tickle Lana.") More elaborate requests and commentaries, each requiring a creative use of a set grammatical form, have been developed subsequently.

Lana monitors her sentences on a computer display, and erases those with grammatical errors. Once, in the midst of Lana's construction of an elaborate sentence, her trainer mischievously and repeatedly interposed, from his separate computer console, a word that made nonsense of Lana's sentence. She gazed at her computer display, spied her trainer at his console, and composed a new sentence: "Please, Tim, leave room." Just as Washoe and Lucy can be said to speak, Lana can be said to write.

Lana at her computer. The overhead bar, just off the top, must be pulled to activate the console. Dispensers for juice, water, bananas, and chocolate candies are near the base of the console.

At an early stage in the development of Washoe's verbal abilities, Jacob Bronowski and a colleague wrote a scientific paper denying the significance of Washoe's use of gestural language because, in the limited data available to Bronowski, Washoe neither inquired nor negated. But later observations showed that Washoe and other chimpanzees were perfectly able both to ask questions and to deny assertions put to them. And it is difficult to see any significant difference in quality between chimpanzee

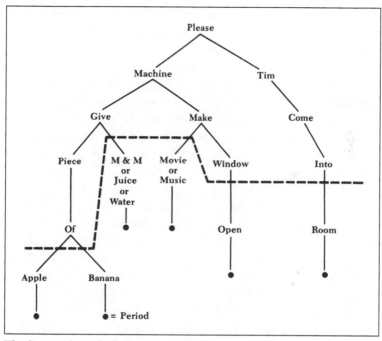

The diagram shows the logic tree required for a number of requests to be communicated. The system is both polite and grammatical: requests must be initiated by a "please" and terminated by a period.

use of gestural language and the use of ordinary speech by children in a manner that we unhesitatingly attribute to intelligence. In reading Bronowski's paper I cannot help but feel that a little pinch of human chauvinism has crept in, an echo of Locke's "Beasts abstract not." In 1949, the American anthropologist Leslie White stated unequivocally: "Human behavior is symbolic behavior; symbolic behavior is human behavior." What would White have made of Washoe, Lucy and Lana?

These findings on chimpanzee language and intelligence have an intriguing bearing on "Rubicon" arguments—the contention that the total brain mass, or at least the ratio of brain to body mass, is a useful index of intelligence. Against this point of view it was once argued that the lower range of the brain masses of microcephalic humans overlaps the upper range of brain masses of adult chimpanzees and gorillas; and yet, it was said, microcephalics have some, although severely impaired, use of language—while the apes have none. But in only relatively few cases are microcephalics capable of human speech. One of the best behavioral descriptions of microcephalics was written by a Russian physician, S. Korsakov, who in 1893 observed a female microcephalic named "Masha." She could understand a very few questions and commands and could occasionally reminisce on her childhood. She sometimes chattered away, but there was little coherence to what she uttered. Korsakov characterized her speech as having "an extreme poverty of logical associations." As an example of her poorly adapted and automaton-like intelligence, Korsakov described her eating habits. When food was present on the table, Masha would eat. But if the food was abruptly removed in the midst of a meal, she would behave as if the meal had ended, thanking those in charge and piously blessing herself. If the food were returned, she would eat again. The pattern apparently was subject to indefinite repetition. My own impression is that Lucy or Washoe would be a far more interesting dinner companion than Masha, and that the comparison of microcephalic humans with normal apes is not inconsistent with some sort of "Rubicon" of intelligence. Of course, both the quality and the quantity of neural connections are probably vital for the sorts of intelligence that we can easily recognize.

Recent experiments performed by James Dewson of the Stanford University School of Medicine and his colleagues give some physiological support to the idea of language centers in the simian neocortex—in particular, like humans, in the left hemisphere.

Monkeys were trained to press a green light when they heard a hiss and a red light when they heard a tone. Some seconds after a sound was heard, the red or the green light would appear at some unpredictable position—different each time—on the control panel. The monkey pressed the appropriate light and, in the case of a correct guess, was rewarded with a pellet of food. Then the time interval between hearing the sound and seeing the light was increased up to twenty seconds. In order to be rewarded, the monkeys now had to remember for twenty seconds which noise they had heard. Dewson's team then surgically excised part of the so-called auditory association cortex from the left hemisphere of the neocortex in the temporal lobe. When retested, the monkeys had very poor recall of which sound they were then hearing. After less than a second they could not recall whether it was a hiss or a tone. The removal of a comparable part of the temporal lobe from the right hemisphere produced no effect whatever on this task. "It looks," Dewson was reported to say, "as if we removed the structure in the monkeys' brains that may be analogous to human language centers." Similar studies on rhesus monkeys, but using visual rather than auditory stimuli, seem to show no evidence of a difference between the hemispheres of the neocortex.

Because adult chimpanzees are generally thought (at least by zookeepers) to be too dangerous to retain in a home or home environment, Washoe and other verbally accomplished chimpanzees have been involuntarily "retired" soon after reaching puberty. Thus we do not yet have experience with the adult language abilities of monkeys and apes. One of the most intriguing questions is whether a verbally accomplished chimpanzee mother will be able to communicate language to her offspring. It seems very likely that this should be possible and that a community of chimps initially competent in gestural language could pass down the language to subsequent generations.

Where such communication is essential for survival, there is already some evidence that apes transmit extragenetic or cultural information. Jane Goodall observed baby chimps in the wild emulating the behavior of their mothers and learning the reasonably complex task of finding an appropriate twig and using it to prod into a termite's nest so as to acquire some of these tasty delicacies.

Differences in group behavior—something that it is very tempting to call cultural differences—have been reported among chimpanzees, baboons, macaques and many other primates. For example, one group of monkeys may know how to eat bird's eggs, while an adjacent band of precisely the same species may not. Such primates have a few dozen sounds or cries, which are used for intra-group communication, with such meanings as "Flee; here is a predator." But the sound of the cries differs somewhat from group to group: there are regional accents.

An even more striking experiment was performed accidentally by Japanese primatologists attempting to relieve an overpopulation and hunger problem in a community of macaques on an island in south Japan. The anthropologists threw grains of wheat on a sandy beach. Now it is very difficult to separate wheat grains one by one from sand grains; such an effort might even expend more energy than eating the collected wheat would provide. But one brilliant macaque, Imo, perhaps by accident or out of pique, threw handfuls of the mixture into the water. Wheat floats; sand sinks, a fact that Imo clearly noted. Through the sifting process she was able to eat well (on a diet of soggy wheat, to be sure). While older macaques, set in their ways, ignored her, the younger monkeys appeared to grasp the importance of her discovery, and imitated it. In the next generation, the practice was more widespread; today all macaques on the island are competent at water sifting, an example of a cultural tradition among the monkeys.

Earlier studies on Takasakiyama, a mountain in northeast Kyushu inhabited by macaques, show a similar pattern in cultural evolution. Visitors to Takasakiyama threw caramels wrapped in paper to the monkeys—a common practice in Japanese zoos, but

A chimp with a long blade of grass, used as a tool to prod termites out of their nest.

one the Takasakiyama macaques had never before encountered. In the course of play, some young monkeys discovered how to unwrap the caramels and eat them. The habit was passed on successively to their playmates, their mothers, the dominant males (who among the macaques act as babysitters for the very young) and finally to the subadult males, who were at the furthest social remove from the monkey children. The process of acculturation took more than three years. In natural primate communities, the existing nonverbal communications are so rich that there is little pressure for the development of a more elaborate gestural language. But if gestural language were necessary for chimpanzee survival, there can be little doubt that it would be transmitted culturally down through the generations.

I would expect a significant development and elaboration of language in only a few generations if all the chimps unable to communicate were to die or fail to reproduce. Basic English corresponds to about 1,000 words. Chimpanzees are already accomplished in vocabularies exceeding 10 percent of that number. Although a few years ago it would have seemed the most implausible science fiction, it does not appear to me out of the question that, after a few generations in such a verbal chimpanzee community, there might emerge the memoirs of the natural history and mental life of a chimpanzee, published in English or Japanese (with perhaps an "as told to" after the by-line).

If chimpanzees have consciousness, if they are capable of abstractions, do they not have what until now has been described as "human rights"? How smart does a chimpanzee have to be before killing him constitutes murder? What further properties must he show before religious missionaries must consider him worthy of attempts at conversion?

I recently was escorted through a large primate research laboratory by its director. We approached a long corridor lined, to the vanishing point as in a perspective drawing, with caged chimpanzees. They were one, two or three to a cage, and I am sure the accommodations were exemplary as far as such institutions (or for that matter traditional zoos) go. As we approached the nearest cage, its two inmates bared their teeth and with incredible accuracy let fly great sweeping arcs of spittle, fairly drenching the lightweight suit of the facility's director. They then uttered a staccato of short shrieks, which echoed down the corridor to be repeated and amplified by other caged chimps, who had certainly not seen us, until the corridor fairly shook with the screeching and banging and rattling of bars. The director informed me that not only spit is apt to fly in such a situation; and at his urging we retreated.

I was powerfully reminded of those American motion pictures of the 1930s and 40s, set in some vast and dehumanized state or federal penitentiary, in which the prisoners banged their eating utensils against the bars at the appearance of the tryrannical warden. These chimps are healthy and well-fed. If they are "only" animals, if they are beasts which abstract not, then my comparison is a piece of sentimental foolishness. But chimpanzees *can* abstract. Like other mammals, they are capable of strong emotions. They have certainly committed no crimes. I do not claim to have the answer, but I think it is certainly worthwhile to raise the question: Why, exactly, all over the civilized world, in virtually every major city, are apes in prison?

For all we know, occasional viable crosses between humans and chimpanzees are possible.* The natural experiment must have been tried very infrequently, at least recently. If such offspring are ever produced, what will their legal status be? The cognitive abilities of chimpanzees force us, I think, to raise searching questions about the boundaries of the community of beings

---

*Until fairly recently it was thought that humans had forty-eight chromosomes in an ordinary somatic cell. We now know that the correct number is forty-six. Chimps apparently really do have forty-eight chromosomes, and in this case a viable cross of a chimpanzee and a human would in any event be rare.

to which special ethical considerations are due, and can, I hope, help to extend our ethical perspectives downward through the taxa on Earth and upwards to extraterrestrial organisms, if they exist.

It is hard to imagine the emotional significance for chimpanzees of learning language. Perhaps the closest analogy is the discovery of language by intelligent human beings with severe sensory organ impairment. While the depth of understanding, intelligence and sensitivity of Helen Keller, who could neither see, hear nor speak, greatly exceeds that of any chimpanzee, her account of her discovery of language carries some of the feeling tone that this remarkable development in primate languages may convey to the chimpanzee, particularly in a context where language enhances survival or is strongly reinforced.

One day Miss Keller's teacher prepared to take her for a walk:

> She brought me my hat, and I knew I was going out into the warm sunshine. This thought, if a wordless sensation may be called a thought, made me hop and skip with pleasure.
>
> We walked down the path to the well-house, attracted by the fragrance of the honeysuckle with which it was covered. Someone was drawing water and my teacher placed my hand under the spout. As the cool stream gushed over my hand she spelled into the other the word *water*, first slowly, then rapidly. I stood still, my whole attention fixed upon the motion of her fingers. Suddenly I felt a misty consciousness as of something forgotten—a thrill of returning thought; and somehow the mystery of language was revealed to me. I knew then that W-A-T-E-R meant the wonderful cool something that was flowing over my hand. That living word awakened my soul, gave it light, hope, joy, set it free! There were barriers still, it is true, but barriers that in time could be swept away.

I left the well-house eager to learn. Everything had a name, and each name gave birth to a new thought. As we returned into the house, every object which I touched seemed to quiver with life. That was because I saw everything with the strange, new sight that had come to me.

Perhaps the most striking aspect of these three exquisite paragraphs is Helen Keller's own sense that her brain had a latent capability for language, needing only to be introduced to it. This essentially Platonic idea is also, as we have seen, consistent with what is known, from brain lesions, of the physiology of the neocortex; and also with the theoretical conclusions drawn by Noam Chomsky of the Massachusetts Institute of Technology from comparative linguistics and laboratory experiments on learning. In recent years it has become clear that the brains of nonhuman primates are similarly prepared, although probably not quite to the same degree, for the introduction of language.

The long-term significance of teaching language to the other primates is difficult to overestimate. There is an arresting passage in Charles Darwin's *Descent of Man:* "The difference in mind between man and the higher animals, great as it is, certainly is one of degree and not of kind.... If it could be proved that certain high mental powers, such as the formation of general concepts, self-consciousness, et cetera, were absolutely peculiar to man, which seems extremely doubtful, it is not improbable that these qualities are merely the incidental results of other highly-advanced intellectual faculties; and these again mainly the results of the continued use of a perfect language."

This same opinion on the remarkable powers of language and human intercommunication can be found in quite a different place, the Genesis account of the Tower of Babel. God, in a strangely defensive attitude for an omnipotent being, is worried that men intend to build a tower that will reach to heaven. (His attitude is similar to the concern he expresses after Adam eats the apple.) To prevent Mankind from reaching heaven, at least metaphorically, God does not destroy the tower, as, for example, Sodom is destroyed. Instead, he says, "Behold, they are one people, and they have all one language; and this is only the beginning of what they will do; and nothing that they propose to do will now be impossible for them. Come, let us go down, and there confuse their language, that they may not understand one another's speech" (Genesis 11:6–7).

The continued use of a "perfect" language ... What sort of culture, what kind of oral tradition would chimpanzees establish after a few hundred or a few thousand years of communal use of a complex gestural language? And if there were such an isolated continuous chimpanzee community, how would they begin to view the origin of language? Would the Gardners and the workers at the Yerkes Primate Center be remembered dimly as legendary folk heroes or gods of another species? Would there be myths, like those of Prometheus, Thoth, or Oannes, about divine beings who had given the gift of language to the apes? In fact, the instruction of chimpanzees in gestural language distinctly has some of the same emotion tone and religious sense of the (truly fictional) episode in the movie and novel *2001: A Space Odyssey* in which a representative of an advanced extraterrestrial civilization somehow instructs our hominid ancestors.

Perhaps the most striking aspect of this entire subject is that there are nonhuman primates so close to the edge of language, so willing to learn, so entirely competent in its use and inventive in its application once the language is taught. But this raises a curious question: *Why* are they all on the edge? Why are there no nonhuman primates with an *existing* complex gestural language? One possible answer, it seems to me, is that humans have systematically exterminated those other primates who displayed signs of intelligence. (This may have been particularly true of the nonhuman primates who lived in the savannahs; the forests must have offered some protection to chimpanzees and gorillas from the depredations of man.) We may have been the agent of natural selection in suppressing the intellectual competition. I think we may have pushed back the frontiers of intelligence and language ability among the nonhuman primates until their intelligence became just indiscernible. In teaching gestural language to the chimpanzees, we are beginning a belated attempt to make amends.

Very old are we men;
Our dreams are tales
Told in dim Eden …

WALTER DE LA MARE
"All That's Past"

"Well, at any rate it's a great comfort," she said as she stepped under the trees, "after being so hot to get into the—into the—into what?" she went on, rather surprised at not being able to think of the word. "I mean to get under the—under the—under this, you know!" putting her hand on the trunk of the tree. "What does it call itself, I wonder?" … And now, who am I? I will remember, if I can! I'm determined to do it!" But being determined didn't help her much, and all she could say, after a great deal of puzzling, was "L, I know it begins with L!"

LEWIS CARROLL
Alice Through the Looking Glass

Come not between the dragon and his wrath.

WM. SHAKESPEARE
King Lear

… At first
Senseless as beasts I gave men sense, possessed them
    of mind …
In the beginning, seeing, they saw amiss, and
    hearing, heard not, but like phantoms huddled
In dreams, the perplexed story of their days
Confounded.

AESCHYLUS
Prometheus Bound

CHAPTER VI

TALES
OF DIM EDEN

PROMETHEUS is in a fit of righteous indignation. He has introduced civilization to a befuddled and superstitious mankind, and
for his pains Zeus has chained him to a rock and set a vulture to
pluck at his liver. In the passage following the above quotation,
Prometheus describes the principal gifts, other than fire, that he
has bestowed on mankind. They are, in order: astronomy; mathematics; writing; the domestication of animals; the invention of
chariots, sailing ships and medicine; and the discovery of divina-
tion by dreams and other methods. The final gift strikes the modern
ear as odd. Along with the account in Genesis of the exile from
Eden, *Prometheus Bound* seems to be one of the major works in
Western literature that present a viable allegory of the evolution
of man—although in this case concentrating much more on the
"evolver" than on the evolved. "Prometheus" is Greek for "foresight," that quality claimed to reside in the frontal lobes of the
neocortex; and foresight and anxiety are both present in Aeschylus' character portrait.

What is the connection between dreams and the evolution of man? Aeschylus is perhaps saying that our prehuman ancestors lived their waking lives in a state akin to our dreaming lives; and that one of the principal benefits of the development of human intelligence is our ability to understand the true nature and import of dreams.

There are, it seems, three principal states of mind in human beings: waking, sleeping and dreaming. An electroencephalograph, which detects brain waves, records quite distinct patterns of electrical activity in the brain during these three states.* Brain waves represent very small currents and voltages produced by the electrical circuitry of the brain. Typical strengths of such brain-wave signals are measured in microvolts. Typical frequencies are between 1 and about 20 Hertz (or cycles per second)—less than the familiar 60 cycles per second frequency of alternating currents in electrical outlets in North America.

But what is sleep good for? There is no doubt that if we stay up too long the body generates neurochemicals that literally force us to go to sleep. Sleep-deprived animals generate such molecules in their cerebrospinal fluid, and the cerebrospinal fluid of sleep-deprived animals induces sleep when injected into other animals who are perfectly wide awake. There must, then, be a very powerful reason for sleep.

The conventional answer of physiology and folk medicine alike is that sleep has a restorative effect; it is an opportunity for the body

---

*The electroencephalograph (EEG) was invented by a German psychologist, Hans Berger, whose fundamental interest in the matter seems to have been telepathy. And, indeed, it can be used for a kind of radio telepathy; human beings have the capability to turn particular brain waves—for example, the alpha rhythm—on and off at will, although this feat requires a little training. With such training, an individual attached to an electroencephalograph and a radio transmitter could, in principle, send quite complex messages in a kind of alpha wave Morse code, merely by thinking them in the right way; and it is just possible that this method might have some practical use, such as permitting patients immobilized by severe stroke to communicate. For historical reasons, non-dreaming sleep is electroencephalographically characterized as "slow wave sleep," and the dream state as "paradoxical sleep."

to perform mental and physical housekeeping away from the needs of daily living. But the actual evidence for this view, apart from its common-sense plausibility, seems to be sparse. Furthermore, there are some worrisome aspects about the contention. For example, an animal is exceptionally vulnerable when sleeping. Granted that most animals sleep in nests, caves, holes in trees or logs or otherwise recessed or camouflaged locations. Even so, their helplessness while asleep remains high. Our nocturnal vulnerability is very evident; the Greeks recognized Morpheus and Thanatos, the gods of sleep and death, as brothers.

Unless there is some exceptionally strong biological necessity for sleep, natural selection would have evolved beasts that sleep not. While there are some animals—the two-toed sloth, the armadillo, the opossum, and the bat—that, at least in states of seasonal torpor, sleep nineteen and twenty hours a day, there are others—the common shrew and Dall's porpoise—that are said to sleep very little. There are also human beings who require only one to three hours of sleep a night. They take second and third jobs, putter around at night while their spouses sink into exhaustion, and otherwise seem to lead full, alert and constructive lives. Family histories suggest that these predispositions are hereditary. In one case, both a man and

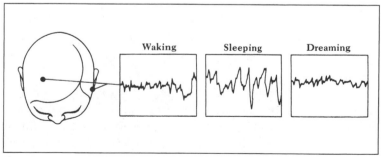

The distinctive EEG patterns of a normal human being while awake, asleep, and dreaming.

his little daughter are afflicted with this blessing or curse, much to the groggy consternation of his wife, who has since divorced him for a novel incompatibility. He retained custody of the daughter. Such examples suggest that the hypothesis of the recuperative function of sleep is at best not the whole story.

Yet sleep is very ancient. In the electroencephalographic sense we share it with all the primates and almost all the other mammals and birds: it may extend back as far as the reptiles. Temporal-lobe epilepsy and its accompanying state of unconscious automatic behavior can be induced in some people by spontaneous electrical stimulation of the amygdala, deep below the temporal lobe, at frequencies of a few cycles per second (a few Hertz). Seizures not very different from sleep have been reported when an epileptic patient is driving in an automobile near sunset or sunrise with a picket fence between him and the sun: at a certain speed the pickets intercept the sun at just the critical rate to produce a flicker at the resonant frequency for initiating such seizures. The circadian rhythm, the daily cycling of physiological function, is known to go back at least to animals as humble as mollusks. Since a state in some respects resembling dreaming can be induced by electrical stimulation of other limbic regions below the temporal lobe, as described below, centers that initiate both sleep and dreams may not be far apart in the recesses of the brain.

There is some recent evidence that the two types of sleep, dreaming and dreamless, depend on the lifestyle of the animal. Truett Allison and Domenic Ciccheti of Yale University have found that predators are statistically much more likely to dream than prey, which are in turn much more likely to experience dreamless sleep. These studies are all of mammals and apply only to differences between, not within, species. In dream sleep, the animal is powerfully immobilized and remarkably unresponsive to external stimuli. Dreamless sleep is much shallower, and we have all witnessed cats or dogs cocking their ears to a sound when apparently fast asleep. It is also commonly held that when sleeping dogs move their legs in a kind of running pattern, they are dreaming of the hunt. The fact that deep dream sleep is rare among prey today seems clearly to be a product of natural selection. But organisms that are largely prey today may have had ancestors that were predators, and vice versa. Moreover, predators are generally organisms with larger absolute brain mass and ratio of brain to body mass than their prey. It makes sense that today, when sleep is highly evolved, the stupid animals are less frequently immobilized by deep sleep than the smart ones. But why should they sleep deeply at all? Why should a state of such deep immobilization ever have evolved?

Perhaps one useful hint about the original function of sleep is to be found in the fact that dolphins and whales and aquatic mammals in general seem to sleep very little. There is, by and large, no place to hide in the ocean. Could it be that, rather than increasing an animal's vulnerability, the function of sleep is to *decrease* it? Wilse Webb of the University of Florida and Ray Meddis of London University have suggested this to be the case. The sleeping style of each organism is exquisitely adapted to the ecology of the animal. It is conceivable that animals who are too stupid to be quiet on their own initiative are, during periods of high risk, immobilized by the implacable arm of sleep. The point seems particularly clear for the young of predatory animals; not only are baby tigers covered with a superbly effective protective coloration, they also sleep a great deal. This is an interesting notion and probably at least partly true. It does not explain everything. Why do lions, who have few natural enemies, sleep? This question is not a very damaging objection because lions may have evolved from animals that were not the kings of beasts. Likewise, adolescent gorillas, who have little to fear, nevertheless construct nests each night— perhaps because they evolved from more vulnerable predecessors. Or perhaps, once, the ancestors of lions and gorillas feared still more formidable predators.

The immobilization hypothesis seems particularly apt in light of the evolution of mammals, who arose in an epoch dominated by

A nest of *Protoceratops* eggs from the Cretaceous of the Mongolian People's Republic.

hissing, thundering and altogether nightmarish reptiles. But nearly all reptiles are cold-blooded* and, except in the tropics, are forced into nocturnal immobility. Mammals are warm-blooded and able to function at night. The nontropical nocturnal ecological niches may have been almost untenanted in the Triassic Period, some two hundred million years ago. Indeed, Harry Jerison has suggested that the evolution of mammals was accompanied by the development of then extremely sophisticated (and now commonplace) versions of hearing and smell, senses for perceiving distances and objects at night; and that the limbic system evolved from the necessity of

---

*Robert Bakker, a paleontologist at Harvard University, suggests that at least some dinosaurs were significantly warm-blooded; even so, it seems likely that they were not as insensitive to diurnal temperature change as mammals are, and that they slowed down substantially at night.

processing the rich array of data from these newly elaborated senses. (A great deal of the visual-information processing in reptiles is done not in the brain but in the retina; the optical processing apparatus in the neocortex was largely a later evolutionary development.)

Perhaps it was essential for the early mammals to be immobilized and hidden during the daylight hours that were ruled by predatory reptiles. I am picturing a late Mesozoic landscape in which the mammals sleep fitfully by day and the reptiles by night. But at night even humble carnivorous protomammals must have posed a real threat to the cold-immobilized reptiles, and particularly to their eggs.

Judged by their endocranial volumes (see figure on page 34), the dinosaurs were, compared to mammals, remarkably stupid. To take some "well-known" examples, *Tyrannosaurus rex* had a brain volume of about 200 cubic centimeters (cc); *Brachiosaurus*, 150 cc; *Triceratops*, 70 cc; *Diplodocus*, 50 cc; *Stegosaurus*, 30 cc. Not one

A reconstruction of baby *Protoceratops* hatching.

A drawing of *Saurornithoides*, a small intelligent dinosaur, here shown catching mammals. Specimens are known from Canada and from the Mongolian People's Republic in the Cretaceous.

approached a chimpanzee in absolute brain mass; *Stegosaurus*, which weighed two metric tons, was probably far more stupid than a rabbit. When the large body weights of the dinosaurs are taken into account, the smallness of their brains becomes even more striking: *Tyrannosaurus* weighed 8 metric tons; *Diplodocus*, 12; and *Brachiosaurus*, 87. The ratio of brain to body weight in *Brachiosaurus* was ten thousand times smaller than that of man. Just as sharks are the largest-brained fish for their body weight, the carnivorous dinosaurs such as *Tyrannosaurus* were relatively larger-brained than such herbivores as *Diplodocus* and *Brachiosaurus*. I am sure that *Tyrannosaurus* was an efficient and terrifying killing machine. But despite their awesome aspect, the dinosaurs look vulnerable to dedicated and intelligent adversaries—such as the early mammals.

Our Mesozoic scene has a curiously vampiric quality with the carnivorous reptiles hunting the smart sleeping mammals by day, and the carnivorous mammals hunting the stupid immobile reptiles by night. While the reptiles buried their eggs, it is unlikely that they actively protected either eggs or young. There are very few accounts of such behavior even in contemporary reptiles, and it is difficult to picture *Tyrannosaurus rex* brooding on a clutch of eggs. For these reasons, the mammals may have won the primordial war of the vampires; at least some paleontologists believe that the demise of the dinosaurs was accelerated by nocturnal predation on reptilian eggs by the early mammals. Two chicken eggs* for breakfast may be all—at least on the surface—that is left of this ancient mammalian cuisine.

The most intelligent of the dinosaurs by the criterion of brain to body mass are the *Saurornithoides*, whose brain mass was typically about 50 grams to a body mass of about 50 kilograms, placing them near the ostrich in the figure on page 35. Indeed, they resembled ostriches. It might be very illuminating to examine fossil endocasts of their braincases. They probably hunted small animals for food and used the four fingers of their handlike appendages for many different tasks. (See illustration above.)

They are interesting beasts to speculate about. If the dinosaurs had not all been mysteriously extinguished some sixty-five million years ago, would the *Saurornithoides* have continued to evolve into increasingly intelligent forms? Would they have learned to hunt large mammals collectively and thus perhaps have prevented the great proliferation of mammals that followed the end of the Mesozoic Age? If it had not been for the extinction of the dinosaurs, would the dominant life forms on Earth today be descendants of *Saurornithoides*, writing and reading books, speculating on what would have

---

*In fact, the birds are almost certainly the principal living descendants of the dinosaurs.

happened had the mammals prevailed? Would the dominant forms think that base 8 arithmetic was quite natural, but base 10 a frill taught only in the "New Math"?

A great deal of what we consider important about the last few tens of millions of years of Earth's history seems to hinge on the extinction of the dinosaurs. There are literally dozens of scientific hypotheses that attempt to explain this event, which appears to have been remarkably rapid and thorough for both land and water forms. All the explanations proposed seem to be only partly satisfactory. They range from massive climatic change to mammalian predation to the extinction of a plant with apparent laxative properties, in which case the dinosaurs died of constipation.

One of the most interesting and promising hypotheses, first suggested by I. S. Shklovskii of the Institute for Cosmic Research, Soviet Academy of Sciences, Moscow, is that the dinosaurs died because of a nearby supernova event—the explosion of a dying star some tens of light-years away, which resulted in an immense flux of high energy charged particles that entered our atmosphere, changed its properties, and, perhaps by destroying the atmospheric ozone, let in lethal quantities of solar ultraviolet radiation. Nocturnal animals, such as the mammals of the time, and deep-sea animals, such as

A reconstruction of a Cretaceous landscape in a swampy region in Western Canada. Most of the dinosaurs shown are bipedal and herbivorous. So far as we know, all these forms became extinct shortly thereafter.

fish, could have survived this higher ultraviolet intensity; but daytime animals that lived on land or near the surface of the waters would have been preferentially destroyed. Such a disaster would be aptly named—the word itself means "bad star."

If this sequence of events is correct, the major course of biological evolution on the Earth in the last sixty-five million years, and indeed the very existence of human beings, can be traced to the death of a distant sun. Perhaps other planets circled that star; perhaps one of those planets enjoyed a thriving biology tortuously evolved over billions of years. The supernova explosion would surely have extinguished all life on that planet and probably even driven its atmosphere into space. Do we owe our existence to a mighty stellar catastrophe that elsewhere destroyed biospheres and worlds?

After the extinction of the dinosaurs, mammals moved into daytime ecological niches. The primate fear of the dark must be a comparatively recent development. Washburn has reported that infant baboons and other young primates appear to be born with only three inborn fears—of falling, snakes, and the dark—corresponding respectively to the dangers posed by Newtonian gravitation to tree-dwellers, by our ancient enemies the reptiles, and by mammalian nocturnal predators, which must have been particularly terrifying for the visually oriented primates.

If the vampiric hypothesis is true—and it is at best a likely hypothesis—the function of sleep is built deeply into the mammalian brain; from earliest mammalian times, sleep played a fundamental role in survival. Since for primitive mammals sleepless nights would have been more dangerous for the survival of the taxon than sexless nights, sleep should be a more powerful drive than sex—which, at least in most of us, it seems to be. But eventually mammals evolved to a point where sleep could be modified by changed circumstances. With the extinction of the dinosaurs, daylight suddenly became a benevolent environment for the mammals. Daytime immobilization was no longer compulsory, and a

Varanus komodoensis, the Komodo dragon, Komodo Island, Indonesia.

wide variety of sleep patterns slowly developed, including the contemporary correlation of mammalian predators with extensive dreaming and mammalian prey with a more watchful dreamless sleep. Perhaps those people who can do with only a few hours' sleep a night are the harbingers of a new human adaptation that will take full advantage of the twenty-four hours of the day. I, for one, freely confess envy for such an adaptation.

These conjectures on the origins of the mammals constitute a kind of scientific myth: they may have some germ of truth in them, but they are unlikely to be the whole story. That scientific myths make contact with more ancient myths may or may not be a coincidence. It is entirely possible that we are able to invent scientific myths only because we have previously been exposed to the other sort. Nevertheless, I cannot resist connecting this account of the origin of mammals with another curious aspect of the Genesis myth of the exile from Eden. Because it is a reptile, of course, that offers the fruit of the knowledge of good and evil—abstract and moral neocortical functions—to Adam and Eve.

There are today a few remaining large reptiles on Earth, the most striking of which is the Komodo dragon of Indonesia: cold-blooded, not very bright, but a predator exhibiting a chilling fixity of purpose. With immense patience, it will stalk a sleeping deer or boar, then suddenly slash a hind leg and hang on until the prey bleeds to death. Prey is tracked by scent, and a hunting dragon lumbers and sashays, head down, its forked tongue flicking over the ground for chemical traces. The largest adults weight about 135 kilograms (300 pounds), are three meters (about 10 feet) long and live perhaps to be centenarians. To protect its eggs, the dragon digs trenches from two to as much as nine meters (almost 30 feet) deep—probably a defense against egg-eating mammals (and themselves: Adults are known occasionally to stalk a nest-hole, waiting for the newly hatched young to emerge and provide a little delicacy for lunch). As another clear adaptation to predators, the dragon hatchlings live in trees.

The remarkable elaboration of these adaptations shows clearly that dragons are in trouble on the planet Earth. The Komodo dragon lives in the wild only in the Lesser Sunda Islands.* There are only about 2,000 of them left. The obscurity of their locale immediately

---

*It is in the Greater Sunda Islands—more specifically Java—that the first specimen of *Homo erectus*, with an endocranial volume of almost 1,000 cc, was found by E. Dubois in 1891.

PHOTO ALINARI

St. George slaying the Dragon, by Donatello from the Chiesa di Or San Michele, Florence.

suggests that dragons are near extinction because of mammalian, chiefly human, predation, a conclusion borne out by their history over the last two centuries. All dragons with less extreme adaptations or less remote habitats are dead. I even wonder whether the systematic separation of brain mass for a given body mass between mammals and reptiles (see chart on page 34) might not be the result of a systematic extinction of bright dragons by mammalian predators. In any case, it is very likely that the population of large reptiles has been declining steadily since the end of the Mesozoic Age, and that there were many more of them even one or two thousand years ago than there are today.

The pervasiveness of dragon myths in the folk legends of many cultures is probably no accident.* The implacable mutual hostility

---

*Curiously, the first representative skull of Peking man—the *Homo erectus* whose remains are clearly associated with the use of fire—was discovered by Weng Chung Pei late in 1929 in Sinkiang Province, China, in a place called the Mountain of Dragons.

between man and dragon, as exemplified in the myth of St. George, is strongest in the West. (In chapter 3 of the Book of Genesis, God ordains an eternal enmity between reptiles and humans.) But it is not a Western anomaly. It is a worldwide phenomenon. Is it only an accident that the common human sounds commanding silence or attracting attention seem strangely imitative of the hissing of reptiles? Is it possible that dragons posed a problem for our protohuman ancestors of a few million years ago, and that the terror they evoked and the deaths they caused helped bring about the evolution of human intelligence? Or does the metaphor of the serpent refer to the use of the aggressive and ritualistic reptilian component of our brain in the further evolution of the neocortex? With one exception, the Genesis account of the temptation by a reptile in Eden is the only instance in the Bible of humans understanding the language of animals. When we feared the dragons, were we fearing a part of ourselves? One way or another, there were dragons in Eden.

The Temptation by a man-serpent and the expulsion from Eden. Michaelangelo, from the ceiling of the Sistine Chapel.

The most recent dinosaur fossil is dated at about sixty million years ago. The family of man (but not the genus *Homo*) is some tens of millions of years old. Could there have been manlike creatures who actually encountered *Tyrannosaurus rex?* Could there have been dinosaurs that escaped the extinctions in the late Cretaceous Period? Could the pervasive dreams and common fears of "monsters," which children develop shortly after they are able to talk, be evolutionary vestiges of quite adaptive—baboonlike—responses to dragons and owls?*

What functions do dreams serve today? One view, published in a reputable scientific paper, holds that the function of dreams is to wake us up a little, every now and then, to see if anyone is about to eat us. But dreams occupy such a relatively small part of normal sleep that this explanation does not seem very compelling. Moreover, as we have seen, the evidence points just the other way: today it is the mammalian predators, not the mammalian prey, who characteristically have dream-filled sleep. Much more plausible is the computer-based explanation that dreams are a spillover from the unconscious processing of the day's experience, from the brain's decision on how much of the daily events temporarily stored in a kind of buffer to emplace in long-term memory. The events of yesterday frequently run through my dreams; the events of two days ago, much more rarely. However, the buffer-dumping model seems unlikely to be the whole story, because it does not explain the disguises that are so characteristic of the symbolic language of dreams, a point first stressed by Freud. It also does not explain the powerful affect or emotions of dreams; I believe there are many people who have

---

*Since writing this passage I have discovered that Darwin expressed a similar thought: "May we not suspect that the vague but very real fears of children, which are quite independent of experience, are inherited effects of real dangers and abject superstitions during ancient savage times? It is quite conformable with what we know of the transmission of formerly well-developed characters, that they should appear at an early period of life, and afterwards disappear"—like gill slits in human embryology.

been far more thoroughly frightened by their dreams than by anything they have ever experienced while awake.

The buffer-dumping and memory-storage functions of dreams have some interesting social implications. The American psychiatrist Ernest Hartmann of Tufts University has provided anecdotal but reasonably persuasive evidence that people who are engaged in intellectual activities during the day, especially unfamiliar intellectual activities, require more sleep at night, while, by and large, those engaged in mainly repetitive and intellectually unchallenging tasks are able to do with much less sleep. However, in part for reasons of organizational convenience, modern societies are structured as if all humans had the same sleep requirements; and in many parts of the world there is a satisfying sense of moral rectitude in rising early. The amount of sleep required for buffer dumping would then depend on how much we have both thought and experienced since the last sleep period. (There is no evidence that the causality runs backwards: people drugged with phenobarbital are not reported, during interstitial waking periods, to perform unusual intellectual accomplishments.) In this respect it would be interesting to examine individuals with very low sleep needs to determine whether the fraction of sleep time they spend dreaming is larger than it is for those with normal sleep requirements, and to determine whether their amount of sleep and dream time increases with the quality and quantity of their learning experiences while awake.

Michel Jouvet, a French neurologist at the University of Lyons, has found that dream sleep is triggered in the pons, which, while it resides in the hindbrain, is a late and essentially mammalian evolutionary development. On the other hand, Penfield has found that electrical stimulation deep into and below the temporal lobe in the neocortex and limbic complex can produce a waking state in epileptics very similar to that of dreams denuded of their symbolic and fantastic aspects. It can also induce the *déjà vu* experience. Much of dream affect, including fear, can also be induced by such electrical stimulation.

I once had a dream that will tantalize me forever. I dreamt I was idly thumbing through a thick history text. I could tell from the illustrations that the work was moving slowly, in the usual manner of such textbooks, through the centuries: classical times, Middle Ages, Renaissance and so on, gradually approaching the modern era. But then *there* was World War II with about two hundred pages left. With mounting excitement I worked my way more deeply into the work until I was sure that I had passed my own time. It was a history book that included the future—like turning the December 31 page of the Cosmic Calendar and finding a fully detailed January 1. Breathlessly I attempted literally to read the future. But it was impossible. I could make out individual words. I could even discern the serifs on the individual characters. But I could not put the letters together into words or words together into sentences. I was alexic.

Perhaps this is simply a metaphor of the unpredictability of the future. But my invariable dream experience is that I am unable to read. I can recognize, for example, a stop sign by its color and its octagonal shape, but I cannot read the word STOP, although I know it is there. I have the impression of understanding the meaning of a page of type, but not by reading it word by word or sentence by sentence. I cannot reliably perform even simple arithmetic operations in the dream state. I make a variety of verbal confusions of no apparent symbolic significance, like mixing up Schumann and Schubert. I am a little aphasic and entirely alexic. Not everyone I know has the same cognitive impairment in the dream state, but people often have some impairment. (Incidentally, individuals who are blind from birth have auditory, not visual dreams.) The neocortex is by no means altogether turned off in the dream state, but it certainly seems to suffer important malfunctions.

The seeming fact that mammals and birds both dream while their common ancestor, the reptiles, do not is surely noteworthy. Major evolution beyond the reptiles has been accompanied by and perhaps requires dreams. The electrically distinctive sleep of birds is episodic and brief. If they dream, they dream for only about a second at a time. But birds are, in an evolutionary sense, much closer to reptiles than mammals are. If we knew only about mammals, the argument would be more shaky; but when both major taxonomic groups that have evolved from the reptiles find themselves compelled to dream, we must take the coincidence seriously. Why should an animal that has evolved from a reptile have to dream while other animals do not? Could it be because the reptilian brain is still present and functioning?

It is extremely rare in the dream state that we bring ourselves up short and say, "This is only a dream." By and large we invest the dream content with reality. There are no rules of internal consistency that dreams are required to follow. The dream is a world of magic and ritual, passion and anger, but very rarely of skepticism and reason. In the metaphor of the triune brain, dreams are partially a function of the R-complex and the limbic cortex, but not of the rational part of the neocortex.

Experiments suggest that as the night wears on our dreams engage increasingly earlier material from our past, reaching back to childhood and infancy. At the same time the primary process and emotional content of the dream also increase. We are much more likely to dream of the passions of the cradle just before awakening than just after falling asleep. This looks very much as if the integration of the day's experience into our memory, the forging of new neural links, is either an easier or a more urgent task. As the night wears on and this function is completed, the more affecting dreams, the more bizarre material, the fears and lusts and other powerful emotions of the dream material emerge. Late at night, when it is very still and the obligatory daily dreams have been dreamt, the gazelles and the dragons begin to stir.

One of the most significant tools in studying the dream state was developed by William Dement, a Stanford University psychiatrist, who is as sane as it is possible for a human being to be, but who bears an exceedingly interesting name for a man of his profession. The dream state is accompanied by rapid eye movements (REM), which can be detected by electrodes taped lightly over the eyelids in sleep, and by a particular brain wave pattern on the EEG. Dement has found that everyone dreams many times each night. On awakening, an individual in the midst of REM sleep will usually remember his dream. Even people who claim never to dream have been discovered by REM and EEG criteria to dream as much as anyone else; and, when awakened at appropriate times, they admit with some surprise to having dreamt. The human brain is in a distinct physiological state while dreaming, and we dream rather often. While perhaps 20 percent of the subjects awakened during REM sleep do not recall their dreams, and some perhaps 10 percent of subjects awakened during non-REM sleep report dreams, we will, for convenience, identify REM and accompanying EEG patterns with the dream state.

There is some evidence that dreaming is necessary. When people or other mammals are deprived of REM sleep (by awakening them as soon as the characteristic REM and EEG dream patterns emerge), the number of initiations of the dream state per night goes up, and, in severe cases, daytime hallucinations—that is, waking dreams—occur. I have mentioned that the REM and EEG patterns of dreams are brief in birds and absent in reptiles. Dreams seem to be primarily a mammalian function. What is more, dream sleep is most vigorously engaged in by human beings in the early postnatal period. Aristotle stated quite positively that infants do not dream at all. On the contrary, we find that they may be dreaming most of the time. Full-term newborn babies spend more than half their sleep time in the REM dream state. In infants born a few weeks premature, the dream time is three-quarters or more of the total sleep time. Earlier in its intrauterine existence, the fetus may be dreaming all the time. (Indeed, newborn kittens are observed to spend all of their sleep time in the REM stage.) Recapitulation would then suggest that dreaming is an evolutionarily early and basic mammalian function.

There is another connection between infancy and dreams: both are followed by amnesia. When we emerge from either state, we have great difficulty remembering what we have experienced. In both cases, I would suggest, the left hemisphere of the neocortex, which is responsible for analytic recollection, has been functioning ineffectively. An alternative explanation is that in both dreams and early childhood we experience a kind of traumatic amnesia: The experiences are too painful to remember. But many dreams we forget are very pleasant, and it is difficult to believe that infancy is *that* unpleasant. Also some children seem capable of remembering extremely early experiences. Memories of events late in the first year of life are not extremely rare, and there are possible examples of even earlier recollections. At age three, my son Nicholas was asked for the earliest event he could recall and replied in a hushed tone while staring into middle distance, "It was red, and I was very cold." He was born by Caesarean section. It is probably very unlikely, but I wonder whether this could just possibly be a true birth memory. At any rate, I think it is much more likely that childhood and dream amnesia arise from the fact that in those states our mental lives are determined almost entirely by the R-complex, the limbic system and the right cerebral hemisphere. In earliest childhood, the neocortex is underdeveloped; in amnesia, it is impaired.

There is a striking correlation of penile or clitoral erection with REM sleep, even when the manifest dream content has no overt sexual aspects whatever. In primates, such erections are connected with sex (of course!), aggression and the maintenance of social hierarchies. I think that when we dream there is a part of us engaged in activities rather like those of the squirrel monkeys I saw in Paul MacLean's laboratory. The R-complex is functioning in the dreams of humans; the dragons can be heard, hissing and rasping, and the dinosaurs thunder still.

One excellent test of the merit of scientific ideas is their subsequent validation. A theory is put forward on fragmentary evidence, then an experiment is performed, the outcome of which the proposer of the theory could not know. If the experiment confirms the original idea, this is usually taken as strong support for the theory. Freud held that the great majority, perhaps all, of the "psychic energy" of our primary-process emotions and dream material is sexual in origin. The absolutely essential role of sexual interest in providing for the propagation of the species makes this idea neither as silly nor as depraved as it appeared to many of Freud's Victorian contemporaries. Carl Gustav Jung, for example, held that Freud had severely overstated the primacy of sex in the affairs of the unconscious. But now, three-quarters of a century later, experiments in the laboratories of Dement and other psychologists appear to support Freud. It would, I think, require a very dedicated puritanism to deny some connection between penile or clitoral erection and sex. It seems to follow that sex and dreams are not casually or incidentally connected but rather have deep and fundamental ties—although dreams certainly partake of ritual, aggressive and hierarchical material as well. Particularly considering the state of sexual repression in late-nineteenth-century Viennese society, many of Freud's insights seem hard-won and courageous as well as valid.

Statistical studies have been made of the most common categories of dreams—studies which, at least to some extent, ought to illuminate the nature of dreams. In a survey of the dreams of college students, the following were, in order, the five most frequent types: (1) falling; (2) being pursued or attacked; (3) attempting repeatedly and unsuccessfully to perform a task; (4) various academic learning experiences; and (5) diverse sexual experiences. Number (4) on this list seems of special and particular concern to the group being sampled. The others, while sometimes actually encountered in the lives of undergraduates, are likely to be applicable generally, even to nonstudents.

The fear of falling seems clearly connected with our arboreal origins and is a fear we apparently share with other primates. If you live in a tree, the easiest way to die is simply to forget the danger of falling. The other three categories of most common dreams are particularly interesting because they correspond to aggressive, hierarchical, ritualistic and sexual functions—the realm of the R-complex. Another provocative statistic is that almost half of the people queried reported dreams about snakes, the only nonhuman animal rating a category all to itself in the twenty most common dreams. It is, of course, possible that many snake dreams have a straightforward Freudian interpretation. However, two-thirds of the respondents reported explicitly sexual dreams. Since, according to Washburn, young primates exhibit an untaught fear of snakes, it is easy to wonder whether the dream world does not point directly as well as indirectly to the ancient hostility between reptiles and mammals.

There is one hypothesis that seems to me consistent with all the foregoing facts: The evolution of the limbic system involved a radically new way of viewing the world. The survival of the early mammals depended on intelligence, daytime unobtrusiveness, and devotion to the young. The world as perceived through the R-complex was quite a different world. Because of the accretionary nature of the evolution of the brain, R-complex functions could be utilized or partially bypassed but not ignored. Thus, an inhibition center developed below what in humans is the temporal lobe, to turn off much of the functioning of the reptilian brain; and an activation center evolved in the pons to turn on the R-complex, but harmlessly, during sleep. This view, of course, has some notable points of similarity to Freud's picture of the repression of the id by the superego (or of the unconscious by the conscious), with expressions of the id made most clearly manifest in slips of the tongue, free associations, dreams and the like—that is, during the interstices of superego repression.

With the large-scale development of the neocortex in higher mammals and primates, some neocortical involvement in the dream state developed—a symbolic language is, after all, still a language. (This is related to the different functions of the two hemispheres of the neocortex, described in the following chapter.) But the dream imagery contained significant sexual, aggressive, hierarchical and ritualistic elements. The fantastic material in the dream world may be connected with the near-absence of direct sensory stimulation during dreams. There is very little reality testing in the dream state. The prevalence of dreams in infants would, in this view, be because, in infancy, the analytic part of the neocortex is barely working. The absence of dreams in reptiles would be because there is no repression of the dream state in reptiles; they are, as Aeschylus described our ancestors, "dreaming" in their waking state. I believe this idea can explain the strangeness—that is, the differences from our waking verbal consciousness—of the dream state; its mammalian and human neonatal localization; its physiology; and its pervasiveness in man.

We are descended from reptiles and mammals both. In the daytime repression of the R-complex and in the nighttime stirring of the dream dragons, we may each of us be replaying the hundred-million-year-old warfare between the reptiles and the mammals. Only the times of day of the vampiric hunt have been reversed.

Human beings exhibit enough reptilian behavior as it is. If we gave full rein to the reptilian aspects of our nature, we would clearly have a low survival potential. Because the R-complex is woven so intimately into the fabric of the brain, its functions cannot be entirely avoided for long. Perhaps the dream state permits, in *our* fantasy and *its* reality, the R-complex to function regularly, as if it were still in control.

If this is true, I wonder, after Aeschylus, if the waking state of other mammals is very much like the dream state of humans—where we can recognize *signs*, such as the feel of running water and the smell of honeysuckle, but have an extremely limited repertoire of *symbols* such as words; where we encounter vivid sensory and emotional images and active intuitive understanding, but very little rational analysis; where we are unable to perform tasks requiring extensive concentration; where we experience short attention spans and frequent distractions and, most of all, a very feeble sense of individuality or self, which gives way to a pervading fatalism, a sense of unpredictable buffeting by uncontrollable events. If this is where we have come from, we have come very far.

Lovers and madmen have such seething brains
Such shaping fantasies, that apprehend
More than cool reason ever comprehends.
The lunatic, the lover, and the poet
Are of imagination all compact ...

WM. SHAKESPEARE
A Midsummer Night's Dream

Mere *poets are as sottish as mere drunkards are, who live in a*
continual mist, without seeing or judging anything clearly. A man
should be learned in several sciences, and should have a
reasonable, philosophical, and in some measure a
mathematical head, to be a complete and excellent poet....

JOHN DRYDEN
"Notes and Observations on The Empress of Morocco," 1674

CHAPTER VII

LOVERS
AND MADMEN

BLOODHOUNDS have a widely celebrated ability to track by smell. They are presented with a "trace"—a scrap of clothing belonging to the target, the lost child or the escaped convict—and then, barking, bound joyously and accurately down the trail. Canines and many other hunting animals have such an ability in extremely well-developed form. The original trace contains an olfactory cue, a smell. A smell is merely the perception of a particular variety of molecule—in this case, an organic molecule. For the bloodhound to track, it must be able to sense the difference in smell—in characteristic body molecules—between the target and a bewildering and noisy background of other molecules, some from other humans who have gone the same way (including those organizing the tracking expedition) and some from other animals (including the dog itself). The number of molecules shed by a human being while walking is relatively small. Yet even on a fairly "cold" trail—say, several hours after the disappearance—bloodhounds can track successfully.

This remarkable ability involves extremely sensitive olfactory detection, a function, as we saw earlier, performed well even by insects. But what is most striking about the bloodhound and different from insects is the richness of its discriminative ability, its aptitude in distinguishing among many different smells, each in an immense background of other odors. The bloodhound performs a sophisticated cataloging of molecular structure; it distinguishes the new molecule from a very large library of other molecules previously smelled. What is more, the bloodhound needs only a minute or less to familiarize itself to the smell, which it can then remember for extensive periods of time.

The olfactory recognition of individual molecules is apparently accomplished by individual nasal receptors sensitive to particular functional groups, or parts, of organic molecules. One receptor, for example, may be sensitive to COOH, another to NH_2, and so on. (C stands for carbon, H for hydrogen, O for oxygen and N for nitrogen.) The various appurtenances and projections of the complex molecules apparently adhere to different molecular receptors in the nasal mucosa, and the detectors for all the functional groups combine to put together a kind of collective olfactory image of the molecule. This is an extremely sophisticated sensory system. The most elaborate man-made device of this sort, the gas chromatograph/mass spectrometer, has in general neither the sensitivity nor the discriminative ability of the bloodhound, although substantial progress is being made in this technology. The olfactory system of animals has evolved into its present sophistication because of strong selection pressures. Early detection of mates, predators and prey is a matter of life and death for the species. The sense of smell is very ancient, and, indeed, much of the early evolution above the level of the neural chassis may have been spurred by selection pressure for such molecular detection: the distinctive olfactory bulbs in the brain (see figure on page 46) are among the first components of the neocortex to have developed in the history of life. Indeed, the limbic system was called the "rhinencephalon," the smell-brain, by Herrick.

The sense of smell is not nearly so well developed in humans as in bloodhounds. Despite the massiveness of our brains, our olfactory bulbs are smaller than those of many other animals, and it is clear that smell plays a very minor role in our everyday lives. The average person is able to distinguish relatively few smells. Our verbal descriptions and analytic comprehension of smell, even with only a few odors in our repertoire, is extremely poor. Our response to an odor hardly resembles, in our own perception, the actual three-dimensional structure of the molecule responsible for the smell. Olfaction is a complex cognitive task which we can, within limits, perform—and with considerable accuracy—but which we can describe inadequately at best. And if the bloodhound could speak, I think it would be at a similar loss to describe the details of what it does so supremely well.

Just as smell is the principal means by which dogs and many other animals perceive their surroundings, sight is the primary information channel in humans. We are capable of visual sensitivity and discrimination at least as impressive as the olfactory abilities of the bloodhound. For example, we are able to discriminate among faces. Careful observers can distinguish among tens or even hundreds of thousands of different faces; and the "Identikit," widely used by Interpol and by police forces in the West generally, is capable of reconstructing more than ten billion different faces. The survival value of such an ability, particularly for our ancestors, is quite clear. Yet consider how incapable we are of describing verbally faces that we are perfectly capable of recognizing. Witnesses commonly exhibit a total failure in verbal description of an individual previously encountered, but high accuracy in recognizing the same individual when seen again. And while cases of mistaken identity have certainly occurred, courts of law seem willing to admit the testimony of any adult witness on questions of facial recognition. Consider how easily we can pick, from a vast crowd of faces, a "celebrity"; or how in a dense nonordered list our own name leaps out at us.

Human beings and other animals have very sophisticated high-data-rate perceptual and cognitive abilities that simply bypass the verbal and analytic consciousness that so many of us regard as all of us there is. This other kind of knowing, our nonverbal perceptions and cognitions, is often described as "intuitive." The word does not mean "innate." No one is born with a repertoire of faces implanted in his brain. The word conveys, I think, a diffuse annoyance at our inability to understand how we come by such knowledge. But intuitive knowledge has an extremely long evolutionary history; if we consider the information contained in the genetic material, it goes back to the origin of life. The other of our two modes of knowing—the one that in the West expresses irritation about the existence of intuitive knowledge—is a quite recent evolutionary accretion. Rational thinking that is fully verbal (involving complete sentences, say) is probably only tens or hundreds of thousands of years old. There are many people who are, in their conscious lives, almost entirely rational, and many who are almost entirely intuitive. Each group, with very little appreciation of the reciprocal value of these two kinds of cognitive ability, derides the other: "muddled" and "amoral" are typical adjectives used in the more polite of such exchanges. Why should we have two different, accurate and complementary modes of thinking which are so poorly integrated with each other?

The first evidence that these two modes of thinking are localized in the cerebral cortex has come from the study of brain lesions. Accidents or strokes in the temporal or parietal lobes of the left hemisphere of the neocortex characteristically result in impairment of the ability to read, write, speak and do arithmetic. Comparable lesions in the right hemisphere lead to impairment of three-dimensional vision, pattern recognition, musical ability and holistic reasoning. Facial recognition resides preferentially in the right hemisphere, and those who "never forget a face" are performing pattern recognition on the right side. Injuries to the right parietal lobe, in fact, sometimes result in the inability of a patient to recognize his own face in a mirror or photograph. Such observations strongly suggest that those functions we describe as "rational" live mainly in the left hemisphere, and those we consider "intuitive," mainly in the right.

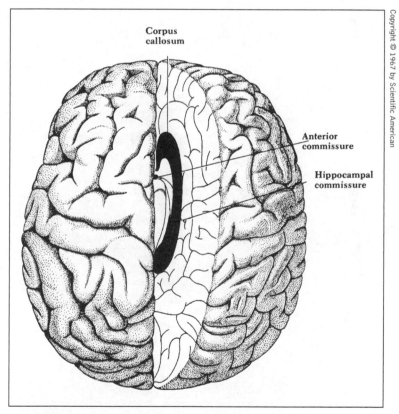

A top view of the human brain, in which the two cerebral hemispheres have been separated by neurosurgeons in a successful attempt to control epileptic seizures. The separation is accomplished principally by cutting the corpus callosum. The more minor connectors of the two hemispheres, the anterior commissure and the hippocampal commissure, are also sometimes cut.

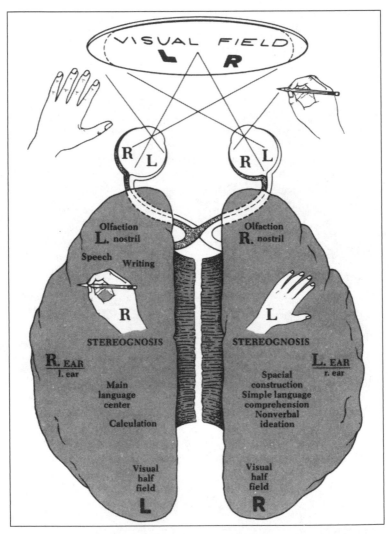

A schematic representation, after Sperry, of the mapping of the outside world onto the two hemispheres of the neocortex. The right and left visual fields are projected, respectively, onto the left and right occipital lobes. Control of the right and left sides of the body is similarly crossed, as is, mainly, hearing. Smells are projected onto the hemispheres on the same side as the nostril doing the smelling.

The most significant recent experiments along these lines have been performed by Roger Sperry and his collaborators at the California Institute of Technology. In an attempt to treat severe cases of *grand mal* epilepsy, where patients suffer from virtually continuous seizures (as frequent as twice an hour, forever), they cut the corpus callosum, the main bundle of neural fibers connecting the left and right hemispheres of the neocortex (see the figure on page 149). The operation was an effort to prevent a kind of neuroelectrical storm in one hemisphere from propagating, far from its focus, into the other. The hope was that at least one of the two postoperative hemispheres would be unaffected by subsequent seizures. The unexpected and welcome result was that the frequency and intensity of the seizures declined dramatically in both hemispheres—as if there had previously been a positive feedback, with the epileptic electrical activity in each hemisphere stimulating the other through the corpus callosum.

Such "split-brain" patients appear, superficially, entirely normal after the surgery. Some report a complete cessation of the vivid dreams they experienced before the operation. The first such patient was unable to speak for a month after the operation, but his aphasia later disappeared. The normal behavior and appearance of split-brain patients in itself suggests that the function of the corpus callosum is subtle. Here is a bundle of two hundred million neural fibers processing something like several billion bits per second between the two cerebral hemispheres. It contains about 2 percent of the total number of neurons in the neocortex. And yet when it is cut, nothing seems to change. I think it is obvious that there must in fact be significant changes, but ones that require a deeper scrutiny.

When we examine an object to our right, both eyes are viewing what is called the right visual field; and to our left, the left visual field. But because of the way the optic nerves are connected, the right visual field is processed in the left hemisphere and the left visual field in the right hemisphere. Likewise, sounds from the right ear are processed primarily in the left hemisphere of the brain and vice

versa, although there is some audio processing on the same side—
for example, sounds from the left ear in the left hemisphere. No
such crossing of function occurs in the more primitive sense of smell,
and an odor detected by the left nostril only is processed exclusively in the left hemisphere. But information sent between the brain
and the limbs *is* crossed. Objects felt by the left hand are perceived
primarily in the right hemisphere, and instructions to the right hand
to write a sentence are processed in the left hemisphere. (See the
figure on page 150.) In 90 percent of human subjects, the centers
for speech are in the left hemisphere.

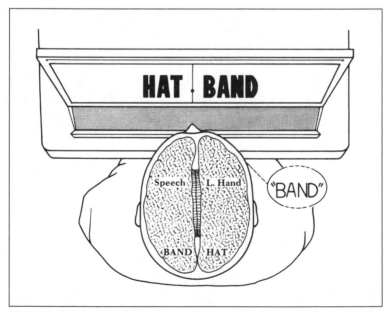

The subject reads and verbally reports only the word flashed to his right visual field.
No association is made, even unconsciously, of the words in left and right visual
fields. After Sperry.

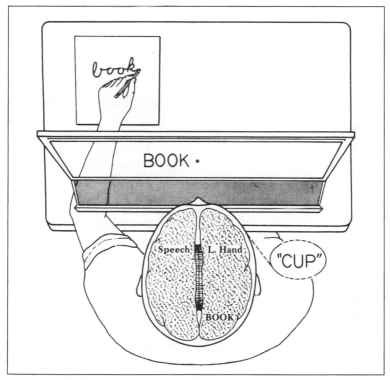

A split-brain patient presented with a word in his left visual field correctly writes (and in script rather than capital letters) the word with the hand out of view. But when the subject is asked what his left hand wrote, he gives a totally incorrect response ("cup"). After Nebes and Sperry.

Sperry and his collaborators have performed an elegant series of experiments in which separate stimuli are presented to the left and right hemispheres of split-brain patients. In a typical experiment, the word *hatband* is flashed on a screen—but *hat* is in the left visual field and *band* in the right visual field. The patient reports that he saw the word *band*, and it is clear that, at least in terms of his ability to communicate verbally, he has no idea that the right hemisphere received a visual impression of the word *hat*. When asked what kind of band it was, the patient might guess: outlaw band, rubber band, jazz band. But when, in comparable experiments, the patient is asked to write what he saw, but with his left hand inside a box, he scrawls the word *hat*. He knows from the motion of his hand that he has written something, but because he cannot see it, there is no way for the information to arrive in the left hemisphere which controls verbal ability. Bewilderingly, he can write, but cannot utter, the answer.

Many other experiments exhibit similar results. In one, the patient is able to feel three-dimensional plastic letters which are out of view with his left hand. The available letters can spell only one correct English word, such as *love* or *cup*, which the patient is able to work out: the right hemisphere has a weak verbal ability, roughly comparable to that in dreams. But after correctly spelling the word, the patient is unable to give any verbal indication of what word he has spelled. It seems clear that in split-brain patients, each hemisphere has scarcely the faintest idea what the other hemisphere has learned.

The geometrical incompetence of the left hemisphere is impressive; it is depicted by the illustration on the opposite page: A right-handed split-brain patient was able to copy simple representations of three-dimensional figures accurately only with his (inexperienced) left hand. The right hemisphere's superiority at geometry seems restricted to manipulative tasks; this dominance does not hold for other sorts of geometrical functions that do not require hand-eye-brain coordination. These manipulative geometrical activities seem to be localized in the right hemisphere's parietal lobe, in a place that, in the left hemisphere, is devoted to language. M. S. Gazzaniga of the State University of New York at Stony Brook suggests that this hemispheric specialization occurs because language is developed in the left hemisphere before the child acquires substantial competence in manipulative skills and geometrical visualization. According to this view, the specialization of the right hemisphere for geometrical competence is a specialization by default—the left hemisphere's competence has been redirected toward language.

Shortly after one of Sperry's most convincing experiments had been completed, he gave a party, so the story goes, to which a famous theoretical physicist with an intact corpus callosum was invited. The physicist, known for his lively sense of humor, sat quietly through

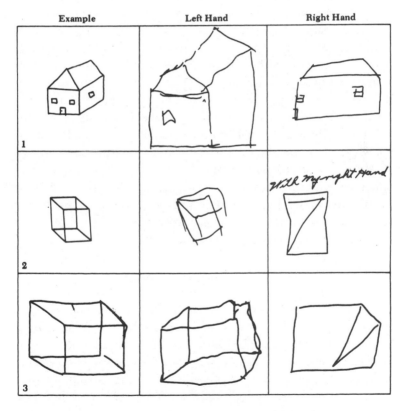

Relative incompetence of the left hemisphere in copying geometrical figures. After Gazzaniga.

the party, listening with interest to Sperry's description of the split-brain findings. The evening passed, the guests trickled away, and Sperry found himself at the door bidding goodbye to the last of them. The physicist extended his right hand, shook Sperry's and told him what a fascinating evening he had had. Then, with a little two-step, he changed the positions of his right and left feet, extended his left hand, and said in a strangled, high-pitched voice, "And I want you to know I had a terrific time too."

When communication between the two cerebral hemispheres is impaired, the patient often finds his own behavior inexplicable, and it is clear that even in "good speaking" the speaker may not know "the truth of the matter." (Compare with the remark on page 2, from the *Phaedrus*.) The relative independence of the two hemispheres is apparent in everyday life. We have already mentioned the difficulty of describing verbally the complex perceptions of the right hemisphere. Many elaborate physical tasks, including athletics, seem to have relatively little left-hemisphere involvement. A well-known "ploy" in tennis, for example, is to ask your opponent exactly where on the racket he places his thumb. It often happens that left-hemisphere attention to this question will, at least for a brief period, destroy his game. A great deal of musical ability is a right-hemisphere function. It is a commonplace that we may memorize a song or a piece of music without having the least ability to write it down in musical notation. In piano, we might describe this by saying that our fingers (but not we) have memorized the piece.

Such memorization can be quite complex. I recently had the pleasure of witnessing the rehearsal of a new piano concerto by a major symphony orchestra. In such rehearsals the conductor does not often start from the beginning and run through to the end. Rather, because of the expense of rehearsal time as well as the competence of the performers, he concentrates on the difficult passages. I was impressed that not only had the soloist memorized the entire piece, she was also able to begin at any requested place in the composition after only a brief glance at the designated measure in the score. This enviable skill is a mixed left- and right-hemisphere function. It is remarkably difficult to memorize a piece of music you have never heard so that you are able to intervene in any measure. In computer terminology, the pianist had random access as opposed to serial access to the composition.

This is a good example of the cooperation between left and right hemispheres in many of the most difficult and highly valued human activities. It is vital not to overestimate the separation of functions on either side of the corpus callosum in a normal human being. The existence of so complex a cabling system as the corpus callosum must mean, it is important to stress again, that interaction of the hemispheres is a vital human function.

In addition to the corpus callosum there is another neural cabling between the left and right hemispheres, which is called the anterior commissure. It is much smaller than the corpus callosum (see figure on page 149), and exists, as the corpus callosum does not, in the brain of the fish. In human split-brain experiments in which the corpus callosum is cut, but not the anterior commissure, olfactory information is invariably transferred between the hemispheres. Occasional transfer of some visual and auditory information through the anterior commissure also seems to occur, but unpredictably from patient to patient. These findings are consistent with anatomy and evolution; the anterior commissure (and the hippocampal commissure; see the figure on page 149) lies deeper than the corpus callosum and transfers information in the limbic cortex and perhaps in other more ancient components of the brain.

Humans exhibit an interesting separation of musical and verbal skills. Patients with lesions of the right temporal lobe or right hemispherectomies are significantly impaired in musical but not in verbal ability—in particular in the recognition and recall of melodies. But their ability to read music is unimpaired. This seems perfectly consistent with the separation of functions described: the memorization and appreciation of music involves the recognition of auditory patterns and a holistic rather than analytic temperament. There is some evidence that poetry is partly a right-hemisphere function; in some cases the patient begins to write poetry for the first time in his life after a lesion in the left hemisphere has left him aphasic. But this would perhaps be, in Dryden's words, "*mere* poetry." Also, the right hemisphere is apparently unable to rhyme.

The separation or lateralization of cortical function was discovered by experiments on brain-damaged individuals. It is, however, important to demonstrate that the conclusions apply to normal humans. Experiments carried out by Gazzaniga present brain-undamaged individuals with a word half in the left and half in the right visual fields, as in split-brain patients, and the reconstruction of the word is monitored. The results indicate that, in the normal brain, the right hemisphere does very little processing of language but instead transmits what it has observed across the corpus callosum to the left hemisphere, where the entire word is put together. Gazzaniga also found a split-brain patient whose right hemisphere was astonishingly competent in language skills: but this patient had experienced a brain pathology in the temporal-parietal region of the left hemisphere at an early age. We have already mentioned the ability of the brain to relocalize functions after injury in the first two years of life, but not thereafter.

Robert Ornstein and David Galin of the Langley Porter Neuropsychiatric Institute in San Francisco claim that as normal people change from analytic to synthetic intellectual activities the EEG activity of the corresponding cerebral hemispheres varies in the predicted way: when a subject is performing mental arithmetic, for example, the right hemisphere exhibits the alpha rhythm characteristic of an "idling" cerebral hemisphere. If this result is confirmed, it would be quite an important finding.

Ornstein offers an interesting analogy to explain why, in the West at least, we have made so much contact with left-hemisphere functions and so little with right. He suggests that our awareness of right-hemisphere function is a little like our ability to see stars in the daytime. The sun is so bright that the stars are invisible, despite the fact that they are just as present in our sky in the daytime as at night. When the sun sets, we are able to perceive the stars. In the same way, the brilliance of our most recent evolutionary accretion, the verbal abilities of the left hemisphere, obscures our awareness of the functions of the intuitive right hemisphere, which in our ancestors must have been the principal means of perceiving the world.*

The left hemisphere processes information sequentially; the right hemisphere simultaneously, accessing several inputs at once. The left hemisphere works in series; the right in parallel. The left hemisphere is something like a digital computer; the right like an analog computer. Sperry suggested that the separation of function in the two hemispheres is the consequence of a "basic incompatibility." Perhaps we are today able to sense directly the operations of the right hemisphere mainly when the left hemisphere has "set"—that is, in dreams.

In the previous chapter, I proposed that a major aspect of the dream state might be the unleashing, at night, of R-complex processes that had been largely repressed by the neocortex during the day. But I mentioned that the important symbolic content of dreams showed significant neocortical involvement, although the frequently reported impairments in reading, writing, arithmetic and verbal recall suffered in dreams were striking.

In addition to the symbolic content of dreams, other aspects of dream imagery point to a neocortical presence in the dream process.

---

*Marijuana is often described as improving our appreciation of and abilities in music, dance, art, pattern and sign recognition and our sensitivity to nonverbal communication. To the best of my knowledge, it is never reported as improving our ability to read and comprehend Ludwig Wittgenstein or Immanuel Kant; to calculate the stresses on bridges; or to compute Laplace transformations. Often the subject has difficulty even in writing down his thoughts coherently. I wonder if, rather than enhancing anything, the cannabinols (the active ingredients in marijuana) simply suppress the left hemisphere and permit the stars to come out. This may also be the objective of the meditative states of many Oriental religions.

For example, I have many times experienced dreams in which the dénouement or critical "plot surprise" was possible only because of clues—apparently unimportant—inserted much earlier into the dream content. The entire plot development of the dream must have been in my mind at the time the dream began. (Incidentally, the time taken for dream events has been shown by Dement to be approximately equal to the time the same events would have taken in real life.) While the content of many dreams seems haphazard, others are remarkably well structured; these dreams have a remarkable resemblance to drama.

We now recognize the very attractive possibility that the left hemisphere of the neocortex is suppressed in the dream state, while the right hemisphere—which has an extensive familiarity with signs but only a halting verbal literacy—is functioning well. It may be that the left hemisphere is not entirely turned off at night but instead is performing tasks that make it inaccessible to consciousness: it is busily engaged in data dumping from the short-term memory buffer, determining what should survive into long-term storage.

There are occasional but reliably reported instances of difficult intellectual problems solved during sleep. Perhaps the most famous is the dream of the German chemist Friedrich Kekulé von Stradonitz. In 1865 the most pressing and puzzling problem in organic structural chemistry was the nature of the benzene molecule. The structure of several simple organic molecules had been deduced from their properties, and all were linear, the constituent atoms being attached to each other in a straight line. According to his own account, Kekulé was dozing on a horse-drawn tram when he had a kind of dream of dancing atoms in linear arrangements. Abruptly the tail of a chain of atoms attached itself to the head and formed a slowly rotating ring. On awakening and recalling this dream fragment, Kekulé realized instantly that the solution to the benzene problem was a hexagonal ring of carbon atoms rather than a straight chain. Observe, however, that this is quintessentially a pattern-recognition exercise and not an analytic activity. It is typical of almost all of the famous creative acts accomplished in the dream state: they are right-hemisphere and not left-hemisphere activities.

The American psychoanalyst Erich Fromm has written: "Must we not expect that, when deprived of the outside world, we regress temporarily to a primitive animal-like unreasonable state of mind? Much can be said in favor of such an assumption, and the view that such a regression is the essential feature of the state of sleep, and thus of dream activity, has been held by many students of dreaming from Plato to Freud." Fromm goes on to point out that we sometimes achieve in the dream state insights that have evaded us when awake. But I believe these insights always have an intuitive or pattern-recognition character. The "animal-like" aspect of the dream state can be understood as the activities of the R-complex and the limbic system, and the occasionally blazing intuitive insight as the activity of the right hemisphere of the neocortex. Both cases occur because in each the repressive functions of the left hemisphere are largely turned off. These right-hemisphere insights Fromm calls "the forgotten language"—and he plausibly argues that they are the common origin of dreams, fairy tales and myths.

In dreams we are sometimes aware that a small portion of us is placidly watching; often, off in a corner of the dream, there is a kind of observer. It is this "watcher" part of our minds that occasionally—sometimes in the midst of a nightmare—will say to us, "This is only a dream." It is the "watcher" who appreciates the dramatic unity of a finely structured dream plot. Most of the time, however, the "watcher" is entirely silent. In psychedelic drug experiences—for example, with marijuana or LSD—the presence of such a "watcher" is commonly reported. LSD experiences may be terrifying in the extreme, and several people have told me that the difference between sanity and insanity in the LSD experience rests entirely on the continued presence of the "watcher," a small, silent portion of the waking consciousness.

In one marijuana experience, my informant became aware of the presence and, in a strange way, the inappropriateness of this silent "watcher," who responds with interest and occasional critical comment to the kaleidoscopic dream imagery of the marijuana experience but is not part of it. "Who *are* you?" my informant silently asked it. "Who wants to know?" it replied, making the experience very like a Sufi or Zen parable. But my informant's question is a deep one. I would suggest the observer is a small part of the critical faculties of the left hemisphere, functioning much more in psychedelic than in dream experiences, but present to a degree in both. However, the ancient query, "Who is it who asks the question?" is still unanswered; perhaps it is another component of the left cerebral hemisphere.

An asymmetry in the temporal lobes in left and right hemispheres of humans and of chimpanzees has been found, with one portion of the left lobe significantly more developed. Human infants are born with this asymmetry (which develops as early as the twenty-ninth week of gestation), thus suggesting a strong genetic predisposition to control speech in the left temporal lobe. (Nevertheless, children with lesions in the left temporal lobe are able, in their first year or two of life, to develop all speech functions in the comparable portion of the right hemisphere with no impairment. At a later age, this replacement is impossible.) Also, lateralization is found in the behavior of young children. They are better able to understand verbal material with the right ear and nonverbal material with the left, a regularity also found in adults. Similarly, infants spend more time on the average looking at objects on their right than at identical objects on their left, and require a louder noise in the left ear than in the right to elicit a response. While no clear asymmetry of these sorts has yet been found in the brains or behavior of apes, Dewson's results (see page 109) suggest that some lateralization may exist in the higher primates; there is no evidence for anatomical asymmetries in the temporal lobes of, say, rhesus monkeys. One would certainly guess that the linguistic abilities of chimpanzees are governed, as in humans, in the left temporal lobe.

The limited inventory of symbolic cries among nonhuman primates seems to be controlled by the limbic system; at least the full vocal repertoire of squirrel and rhesus monkeys can be evoked by electrical stimulation in the limbic system. Human language is controlled in the neocortex. Thus an essential step in human evolution must have been the transfer of control of vocal language from the limbic system to the temporal lobes of the neocortex, a transition from instinctual to learned communication. However, the surprising ability of apes to acquire gestural language and the hint of lateralization in the chimpanzee brain suggest that the acquisition of voluntary symbolic language by primates is not a recent invention. Rather, it goes back many millions of years, consistent with the evidence from endocranial casts for Broca's area in *Homo habilis.*

Lesions in the monkey brain of the neocortical areas responsible for speech in humans fail to impair their instinctual vocalizations. The development of human language must therefore involve an essentially new brain system and not merely a reworking of the machinery for limbic cries and calls. Some experts in human evolution have suggested that the acquisition of language occurred very late—perhaps only in the last few tens of thousands of years—and was connected with the challenges of the last ice age. But the data do not seem to be consistent with this view; moreover, the speech centers of the human brain are so complex that it is very difficult to imagine their evolution in the thousand or so generations since the peak of the most recent glaciation.

The evidence suggests that in our ancestors of some tens of millions of years ago there was a neocortex, but one in which the left and right hemispheres served comparable and redundant functions. Since then, upright posture, the use of tools, and the development of language have mutually advanced one another, a small increment in language ability, for example, permitting the incremental improvement of hand axes, and vice versa. The corresponding brain evolution seems to have proceeded by specializing one of the two hemispheres for analytic thinking.

The original redundancy, by the way, represents prudent computer design. For example, with no knowledge of the neuroanatomy of the cerebral cortex, the engineers who designed the on-board memory of the Viking lander inserted two identical computers, which are identically programmed. But because of their complexity, differences between the computers soon emerged. Before landing on Mars the computers were given an intelligence test (by a smarter computer back on Earth). The dumber brain was then turned off. Perhaps human evolution has proceeded in a similar manner and our highly prized rational and analytical abilities are localized in the "other" brain—the one that was not fully competent to do intuitive thinking. Evolution often uses this strategy. Indeed, the standard evolutionary practice of increasing the amount of genetic information as organisms increase in complexity is accomplished by doubling part of the genetic material and then allowing the slow specialization of function of the redundant set.

Almost without exception all human languages have built into them a polarity, a veer to the right. "Right" is associated with legality, correct behavior, high moral principles, firmness, and masculinity; "left," with weakness, cowardice, diffuseness of purpose, evil, and femininity. In English, for example, we have "rectitude," "rectify," "righteous," "right-hand man," "dexterity," "adroit" (from the French "à droite"), "rights," as in "the rights of man," and the phrase "in his right mind." Even "ambidextrous" means, ultimately, two right hands.

On the other side (literally), we have "sinister" (almost exactly the Latin word for "left"), "gauche" (precisely the French word for "left"), "gawky," "gawk," and "left-handed compliment." The Russian "nalevo" for "left" also means "surreptitious." The Italian "mancino" for "left" signifies "deceitful." There is no "Bill of Lefts."

In one etymology, "left" comes from "lyft," the Anglo-Saxon for weak or worthless. "Right" in the legal sense (as an action in accord with the rules of society) and "right" in the logical sense (as the opposite of erroneous) are also commonplaces in many languages. The political use of right and left seems to date from the moment when a significant lay political force arose as counterpoise to the nobility. The nobles were placed on the king's right and the radical upstarts—the capitalists—on his left. The nobles were to the royal right, of course, because the king himself was a noble; and his right side was the favored position. And in theology as in politics: "At the right hand of God."

Many examples of a connection between "right" and "straight" can be found.* In Mexican Spanish you indicate straight (ahead) by saying "right right"; in Black American English, "right on" is an expression of approval, often for a sentiment eloquently or deftly phrased. "Straight" meaning conventional, correct or proper is a commonplace in colloquial English today. In Russian, right is "*pravo*," a cognate of "*pravda*," which means "true." And in many languages "true" has the additional meaning of "straight" or "accurate," as in "his aim was true."

The Stanford-Binet IQ test makes some effort to examine both left- and right-hemisphere function. For right-hemisphere function there are tests in which the subject is asked to predict the opened configuration of a piece of paper after it is folded several times and a small piece cut out with a pair of scissors; or to estimate the total number of blocks in a stack when some blocks are hidden from view. Although the devisers of the Stanford-Binet test consider such questions of geometric conception to be very useful in determining the "intelligence" of children, they are said to be increasingly less useful in IQ tests of teenagers and adults. There is certainly little room on

---

*I wonder if there is any significance to the fact that Latin, Germanic and Slavic languages, for example, are written left to right, and Semitic languages, right to left. The ancient Greeks wrote in boustrophedon ("as the ox plows"): left to right on one line, right to left on the next.

such examinations for testing intuitive leaps. Unsurprisingly, IQ tests also seem to be powerfully biased toward the left hemisphere.

The vehemence of the prejudices in favor of the left hemisphere and the right hand reminds me of a war in which the side that barely won renames the contending parties and issues, so that future generations will have no difficulty in deciding where prudent loyalty should lie. When Lenin's party was a fairly small splinter group in Russian socialism he named it the Bolshevik party, which in Russian means the majority party. The opposition obligingly, and with awesome stupidity, accepted the designation of Mensheviks, the minority party. In a decade and a half they were. Similarly, in the worldwide associations of the words "right" and "left" there is evidence of a rancorous conflict early in the history of mankind.* What could arouse such powerful emotions?

In combat with weapons which cut or stab—and in such sports as boxing, baseball and tennis—a participant trained in the use of the right hand will find himself at a disadvantage when confronted unexpectedly with a left-hander. Also, a malevolent left-handed swordsman might be able to come quite close to his adversary with his unencumbered right hand appearing as a gesture of disarmament and peace. But these circumstances do not seem to be able to explain the breadth and depth of antipathy to the left hand, nor the extension of right chauvinism to women—traditional noncombatants.

One, perhaps remote, possibility is connected with the unavailability of toilet paper in preindustrial societies. For most of human history, and in many parts of the world today, the empty hand is used

---

*A quite different set of circumstances is revealed by another pair of verbal polar opposites: black and white. Despite English phrases of the sort "as different as black and white," the two words appear to have the same origin. Black comes from the Anglo-Saxon "*blaece*," and white from the Anglo-Saxon "*blac*," which is still active in its cognates "blanch," "blank," "bleak," and the French "*blanc*." Both black and white have as their distinguishing properties the absence of color, and employing the same word for both strikes me as very perceptive of King Arthur's lexicographer.

for personal hygiene after defecation, a fact of life in pretechnological cultures. It does not follow that those who follow this custom enjoy it. Not only is it aesthetically unappealing, it involves a serious risk of transferring disease to others as well as to oneself. The simplest precaution is to greet and to eat with the other hand. Without apparent exception in pretechnological human societies, it is the left hand that is used for such toilet functions and the right for greeting and eating. Occasional lapses from this convention are quite properly viewed with horror. Severe penalties have been visited on small children for breaches of the prevailing handedness conventions; and many older people in the West can still remember a time when there were firm strictures against even reaching for objects with the left hand. I believe this account can explain the virulence against associations with "left" and the defensive self-congratulatory bombast attached to associations with "right" which are commonplace in our right-handed society. The explanation does not, however, explain why the right and left hands were originally chosen for these particular functions. It might be argued that statistically there is one chance in two that toilet functions would be relegated to the left hand. But we would then expect one society in two to be righteous about leftness. In fact, there seem to be no such societies. In a society where most people are right-handed, precision tasks such as eating and fighting would be relegated to the favored hand, leaving by default toilet functions to the side sinister. However, this also does not account for why the society is right-handed. In its most fundamental sense, the explanation must lie elsewhere.

There is no direct connection between the hand you prefer to use for most tasks and the cerebral hemisphere that controls speech, and the majority of left-handers may still have speech centers in the left hemisphere, although this point is in dispute. Nevertheless, the existence of handedness itself is thought to be connected with brain lateralization. Some evidence suggests that left-handers are more likely to have problems with such left-hemisphere functions as reading, writing, speaking and arithmetic; and to be more adept at such

Two robust Australopithecines. These animals may have been predominantly right-handed; the gracile Australopithecines very likely were.

right-hemisphere functions as imagination, pattern recognition and general creativity.* Some data suggest that human beings are *genetically* biased towards right-handedness. For example, the number of ridges on fingerprints of fetuses during the third and fourth months of gestation is larger in the right hand than the left hand, and this preponderance persists throughout fetal life and after birth.

Information on the handedness of the Australopithecines has been obtained from an analysis of fossil baboon skulls fractured with bone or wooden clubs by these early relatives of man. The discoverer of the Australopithecine fossils, Raymond Dart, concluded that about 20 percent of them were left-handed, which is roughly the fraction in modern man. In contrast, while other animals often show strong paw preferences, the favored paw is almost as likely to be left as right.

The left/right distinctions run deep into the past of our species. I wonder if some slight whiff of the battle between the rational and the intuitive, between the two hemispheres of the brain, has not surfaced in the polarity between words for right and left: it is the verbal hemisphere that controls the right side. There may not in fact be more dexterity in the right side; but it certainly has a better press. The left hemisphere seems to feel quite defensive—in a strange way insecure—about the right hemisphere; and, if this is so, verbal criticism of intuitive thinking becomes suspect on the ground of motive. Unfortunately, there is every reason to think that the right hemisphere has comparable misgivings—expressed nonverbally, of course—about the left.

Admitting the validity of both methods of thinking, left hemisphere and right hemisphere, we must ask if they are equally effective

---

*The only left-handed American presidents have apparently been Harry Truman and Gerald Ford. I am not sure whether this is consistent or inconsistent with the proposed (weak) correlation between handedness and hemisphere function. Leonardo da Vinci may be the most illuminating example of the creative genius of left-handers.

and useful in new circumstances. There is no doubt that right-hemisphere intuitive thinking may perceive patterns and connections too difficult for the left hemisphere; but it may also detect patterns where none exist. Skeptical and critical thinking is not a hallmark of the right hemisphere. And unalloyed right-hemisphere doctrines, particularly when they are invented during new and trying circumstances, may be erroneous or paranoid.

Recent experiments by Stuart Dimond, a psychologist at University College, Cardiff in Wales, have employed special contact lenses to show films to the right or left hemisphere only. Of course, the information arriving in one hemisphere in a normal subject can be transmitted via the corpus callosum to the other hemisphere. Subjects were asked to rate a variety of films in terms of emotional content. These experiments showed a remarkable tendency for the right hemisphere to view the world as more unpleasant, hostile, and even disgusting than the left hemisphere. The Cardiff psychologists also found that when both hemispheres are working, our emotional responses are very similar to those of the left hemisphere only. The negativism of the right hemisphere is apparently strongly tempered in everyday life by the more easygoing left hemisphere. But a dark and suspicious emotion tone seems to lurk in the right hemisphere, which may explain some of the antipathy felt by our left hemisphere selves to the "sinister" quality of the left hand and the right hemisphere.

In paranoid thinking a person believes he has detected a conspiracy—that is, a hidden (and malevolent) pattern in the behavior of friends, associates or governments—where in fact no such pattern exists. If there *is* such a conspiracy, the subject may be profoundly anxious, but his thinking is not necessarily paranoid. A famous case involves James Forrestal, the first U.S. Secretary of Defense. At the end of World War II, Forrestal was convinced that Israeli secret agents were following him everywhere. His physicians, equally convinced of the absurdity of this *idée fixe*, diagnosed him as paranoid and confined him to an upper story of Walter Reed Army Hospital, from which he plunged to his death, partly because of inadequate supervision by hospital personnel, overly deferential to one of his exalted rank. Later it was discovered that Forrestal was indeed being followed by Israeli agents who were worried that he might reach a secret understanding with representatives of Arab nations. Forrestal had other problems, but having his valid perception labeled paranoid did not help his condition.

In times of rapid social change there are bound to be conspiracies, both by those in favor of change and by those defending the status quo, the latter more than the former in recent American political history. Detecting conspiracies when there are no conspiracies is a symptom of paranoia; detecting them when they exist is a sign of mental health. An acquaintance of mine says, "In America today, if you're not a little paranoid you're out of your mind." The remark, however, has global applicability.

There is no way to tell whether the patterns extracted by the right hemisphere are real or imagined without subjecting them to left-hemisphere scrutiny. On the other hand, mere critical thinking, without creative and intuitive insights, without the search for new patterns, is sterile and doomed. To solve complex problems in changing circumstances requires the activity of both cerebral hemispheres: the path to the future lies through the corpus callosum.

An example of different behavior arising from different cognitive functions—one example of many—is the familiar human reaction to the sight of blood. Many of us feel queasy or disgusted or even faint at the sight of copious bleeding in someone else. The reason, I think, is clear. We have over the years associated our own bleeding with pain, injury, and a violation of bodily integrity; and we experience a sympathetic or vicarious agony in seeing someone else bleed. We recognize their pain. This is almost certainly the reason that the color red is used to signify danger or stop* in many diverse

*Or down, as in elevator direction lights. Our arboreal ancestors had to be very careful about down.

human societies. (If the oxygen-carrying pigment in our blood were green—which biochemically it could have been—we would, all of us, think green a quite natural index of danger and be amused at the idea of using red.) A trained physician, on the other hand, has a different set of perceptions when faced with blood. What organ is injured? How copious is the bleeding? Is it venous or arterial flow? Should a tourniquet be applied? These are all analytic functions of the left hemisphere. They require more complex and analytic cognitive processes than the simple association: blood equals pain. And they are far more practical. If I were injured, I would much rather be with a competent physician who through long experience has become almost entirely inured to gore than with an utterly sympathetic friend who faints dead away at the sight of blood. The latter may be highly motivated not to wound another person, but the former will be able to help if such a wound occurs. In an ideally structured species, these two quite different attitudes would be present simultaneously in the same individual. And in most of us that is just what has happened. The two modes of thinking are of very different complexity, but they have complementary survival value.

A typical example of the occasional resistance mustered by intuitive thinking against the clear conclusions of analytical thinking is D. H. Lawrence's opinion of the nature of the moon: "It's no use telling me it's a dead rock in the sky! I *know* it's not." Indeed, the moon *is* more than a dead rock in the sky. It is beautiful, it has romantic associations, it raises tides, it may even be the ultimate reason for the timing of the human menstrual cycle. But certainly one of its attributes is that it is a dead rock in the sky. Intuitive thinking does quite well in areas where we have had previous personal or evolutionary experience. But in new areas—such as the nature of celestial objects close up—intuitive reasoning must be diffident in its claims and willing to accommodate to the insights that rational thinking wrests from Nature. By the same token, the processes of rational thought are not ends in themselves but must be perceived in the larger context of human good; the nature and direction of rational and analytical endeavors should be determined in significant part by their ultimate human implications, as revealed through intuitive thinking.

In a way, science might be described as paranoid thinking applied to Nature: we are looking for natural conspiracies, for connections among apparently disparate data. Our objective is to abstract patterns from Nature (right-hemisphere thinking), but many proposed patterns do not in fact correspond to the data. Thus all proposed patterns must be subjected to the sieve of critical analysis (left-hemisphere thinking). The search for patterns without critical analysis, and rigid skepticism without a search for patterns, are the antipodes of incomplete science. The effective pursuit of knowledge requires both functions.

Calculus, Newtonian physics and geometrical optics were all derived by fundamentally geometrical arguments and are today taught and demonstrated largely by analytical arguments: creating the mathematics and physics is more of a right-hemisphere function than teaching it. This is common today as well. Major scientific insights are characteristically intuitive, and equally characteristically described in scientific papers by linear analytical arguments. There is no anomaly in this: it is, rather, just as it should be. The creative act has major right-hemisphere components. But arguments on the validity of the result are largely left-hemisphere functions.

It was an astonishing insight by Albert Einstein, central to the theory of general relativity, that gravitation could be understood by setting the contracted Riemann-Christoffel tensor equal to zero. But this contention was accepted only because one could work out the detailed mathematical consequences of the equation, see where it made predictions different from those of Newtonian gravitation, and then turn to experiment to see which way Nature votes. In three remarkable experiments—the deflection of starlight when passing near the sun; the motion of the orbit of Mercury, the planet nearest to the sun; and the red shift of spectral lines in a strong stellar gravitational field—Nature voted for Einstein. But without these experimental tests, very few physicists would have accepted general relativity. There are many hypotheses in physics of almost comparable brilliance and elegance that have been rejected because they did not survive such a confrontation with experiment. In my view, the human condition would be greatly improved if such confrontations and willingness to reject hypotheses were a regular part of our social, political, economic, religious and cultural lives.

I know of no significant advance in science that did not require major inputs from both cerebral hemispheres. This is not true for art, where apparently there are no experiments by which capable, dedicated and unbiased observers can determine to their mutual satisfaction which works are great. As one of hundreds of examples, I might note that the principal French art critics, journals and museums of the late nineteenth and early twentieth centuries rejected French Impressionism *in toto*; today the same artists are widely held by the same institutions to have produced masterpieces. Perhaps a century hence the pendulum will reverse direction again.

This book itself is an exercise in pattern recognition, an attempt to understand something of the nature and evolution of human intelligence, using clues from a wide variety of sciences and myths. It is in significant part a right-hemisphere activity; and in the course of writing it I was repeatedly awakened in the middle of the night or in the early hours of the morning by the mild exhilaration of a new insight. But whether the insights are genuine—and I expect many of them will require substantial revision—depends on how well my *left* hemisphere has functioned (and also on whether I have retained certain views because I am unaware of the evidence that contradicts them). In writing this book I have been repeatedly struck by its existence as a meta-example: in conception and execution it illustrates its own content.

In the seventeenth century there were two quite distinct ways of describing the connection between mathematical quantities: you could write an algebraic equation or you could draw a curve. René Descartes showed the formal identity of these two views of the mathematical world when he invented analytical geometry, through which algebraic equations can be graphed. (Descartes, incidentally, was also an anatomist concerned about the localization of function in the brain.) Analytical geometry is now a tenth-grade commonplace, but it was a brilliant discovery for the seventeenth century. However, an algebraic equation is an archetypical left-hemisphere construction, while a regular geometrical curve, the pattern in an array of related points, is a characteristic right-hemisphere production. In a certain sense, analytical geometry is the corpus callosum of mathematics. Today a range of doctrines find themselves either in conflict or without mutual interaction. In some important instances, they are left-hemisphere versus right-hemisphere views. The Cartesian connection of apparently unrelated or antithetical doctrines is sorely needed once again.

I think the most significant creative activities of our or any other human culture—legal and ethical systems, art and music, science and technology—were made possible only through the collaborative work of the left and right cerebral hemispheres. These creative acts, even if engaged in rarely or only by a few, have changed us and the world. We might say that human culture is the function of the corpus callosum.

It is the business of the future to be dangerous. . . .
The major advances in civilization are processes that all but wreck
the societies in which they occur.

ALFRED NORTH WHITEHEAD
Adventures in Ideas

The voice of the intellect is a soft one, but it does not rest until it has
gained a hearing. Ultimately, after endless rebuffs, it succeeds.
This is one of the few points in which one may be optimistic about
the future of mankind.

SIGMUND FREUD
The Future of an Illusion

The mind of man is capable of anything—because everything is in it,
all the past as well as all the future.

JOSEPH CONRAD
Heart of Darkness

CHAPTER VIII

THE FUTURE
EVOLUTION
OF THE BRAIN

THE HUMAN BRAIN seems to be in a state of uneasy truce, with occasional skirmishes and rare battles. The existence of brain components with predispositions to certain behavior is not an invitation to fatalism or despair: we have substantial control over the relative importance of each component. Anatomy is not destiny, but it is not irrelevant either. At least some mental illness can be understood in terms of a conflict among the contending neural parties. The mutual repression among the components goes in many directions. We have discussed limbic and neocortical repression of the R-complex, but through society, there may also be R-complex repression of the neocortex, and repression of one cerebral hemisphere by the other.

In general, human societies are not innovative. They are hierarchical and ritualistic. Suggestions for change are greeted with suspicion: they imply an unpleasant future variation in ritual

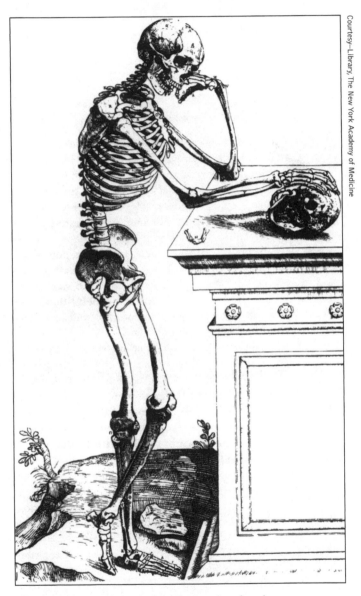

Man ponders himself. By Vesalius, the founder of modern anatomy.

and hierarchy: an exchange of one set of rituals for another, or perhaps for a less structured society with fewer rituals. And yet there are times when societies must change. "The dogmas of the quiet past are inadequate for the stormy present" was Abraham Lincoln's description of this truth. Much of the difficulty in attempting to restructure American and other societies arises from this resistance by groups with vested interests in the status quo. Significant change might require those who are now high in the hierarchy to move downward many steps. This seems to them undesirable and is resisted.

But some change, in fact some significant change, is apparent in Western society—certainly not enough, but more than in almost any other society. Older and more static cultures are much more resistant to change. In Colin Turnbull's book *The Forest People*, there is a poignant description of a crippled Pygmy girl who was provided by visiting anthropologists with a stunning technological innovation, the crutch. Despite the fact that it greatly eased the suffering of the little girl, the adults, including her parents, showed no particular interest in this invention.* There are many other cases of intolerance to novelty in traditional societies; and diverse pertinent examples could be drawn from the lives of such men as Leonardo, Galileo, Desiderius Erasmus, Charles Darwin, or Sigmund Freud.

---

*In defense of the Pygmies, perhaps I should note that a friend of mine who has spent time with them says that for such activities as the patient stalking and hunting of mammals and fish they prepare themselves through marijuana intoxication, which helps to make the long waits, boring to anyone further evolved than a Komodo dragon, at least moderately tolerable. Ganja is, he says, their only cultivated crop. It would be wryly interesting if in human history the cultivation of marijuana led generally to the invention of agriculture, and thereby to civilization. (The marijuana-intoxicated Pygmy, poised patiently for an hour with his fishing spear aloft, is earnestly burlesqued by the beer-sodden riflemen, protectively camouflaged in red plaid, who, stumbling through the nearby woods, terrorize American suburbs each Thanksgiving.)

The traditionalism of societies in a static state is generally adaptive: the cultural forms have been evolved painfully over many generations and are known to serve well. Like mutations, any random change is apt to serve less well. But also like mutations, changes are necessary if adaptation to new environmental circumstances is to be achieved. The tension between these two tendencies marks much of the political conflict of our age. At a time characterized by a rapidly varying external physical and social environment—such as our time—accommodation to and acceptance of change is adaptive; in societies that inhabit static environments, it is not. The hunter/gatherer lifestyles have served mankind well for most of our history, and I think there is unmistakable evidence that we are in a way designed by evolution for such a culture; when we abandon the hunter/gatherer life we abandon the childhood of our species. Hunter/gatherer and high technology cultures are both products of the neocortex. We are now irreversibly set upon the latter path. But it will take some getting used to.

Britain has produced a range of remarkably gifted multidisciplinary scientists and scholars who are sometimes described as polymaths. The group included, in recent times, Bertrand Russell, A. N. Whitehead, J. B. S. Haldane, J. D. Bernal, and Jacob Bronowski. Russell commented that the development of such gifted individuals required a childhood period in which there was little or no pressure for conformity, a time in which the child could develop and pursue his or her own interests no matter how unusual or bizarre. Because of the strong pressures for social conformity both by the government and by peer groups in the United States—and even more so in the Soviet Union, Japan and the People's Republic of China—I think that such countries are producing proportionately fewer polymaths. I also think there is evidence that Britain is in a steep current decline in this respect.

Particularly today, when so many difficult and complex problems face the human species, the development of broad and powerful thinking is desperately needed. There should be a way, consistent

A hunter/gatherer simultaneously stalking prey and educating the young. This life style, which has been characteristic of our species for millions of years, is now almost extinct.

with the democratic ideals espoused by all of these countries, to encourage, in a humane and caring context, the intellectual development of especially promising youngsters. Instead we find, in the instructional and examination systems of most of these countries, an almost reptilian ritualization of the educational process. I sometimes wonder whether the appeal of sex and aggression in contemporary American television and film offerings reflects the fact that the R-complex is well developed in all of us, while many neocortical functions are, partly because of the repressive nature of schools and societies, more rarely expressed, less familiar and insufficiently treasured.

As a consequence of the enormous social and technological changes of the last few centuries, the world is not working well. We do not live in traditional and static societies. But our governments, in resisting change, act as if we did. Unless we destroy ourselves utterly, the future belongs to those societies that, while not ignoring the reptilian and mammalian parts of our being, enable the characteristically human components of our nature to flourish; to those societies that encourage diversity rather than conformity; to those societies willing to invest resources in a variety of social, political, economic and cultural experiments, and prepared to sacrifice short-term advantage for long-term benefit; to those societies that treat new ideas as delicate, fragile and immensely valuable pathways to the future.

A better understanding of the brain may also one day bear on such vexing social issues as the definition of death and the acceptability of abortions. The current ethos in the West seems to be that it is permissible in a good cause to kill nonhuman primates and certainly other mammals; but it is impermissible (for individuals) to kill human beings under similar circumstances. The logical implication is that it is the characteristically human qualities of the human brain that make the difference. In the same way, if substantial parts of the neocortex are functioning, the comatose patient can certainly be said to be alive in a human sense, even if there is major impairment of other physical and neurological functions. On the other hand, a patient otherwise alive but exhibiting no sign of neocortical activity (including the neocortical activities in sleep) might, in a human sense, be described as dead. In many such cases the neocortex has failed irreversibly but the limbic system, R-complex, and lower brainstem are still operative, and such fundamental functions as respiration and blood circulation are unimpaired. I think more work is required on human brain physiology before a well-supported legal definition of death can be generally accepted, but the road to such a definition will very likely take us through considerations of the neocortex as opposed to the other components of the brain.

Similar ideas could help to resolve the great abortion debate flourishing in America in the late 1970s—a controversy marked on both sides by extreme vehemence and a denial of any merit to opposing points of view. At one extreme is the position that a woman has an innate right of "control of her own body," which encompasses, it is said, arranging for the death of a fetus on a variety of grounds including psychological disinclination and economic inability to raise a child. At the other extreme is the existence of a "right to life," the assertion that the killing of even a zygote, a fertilized egg before the first embryonic division, is murder because the zygote has the "potential" to become a human being. I realize that in an issue so emotionally charged any proposed solution is unlikely to receive plaudits from the partisans of either extreme, and sometimes our hearts and our heads lead us to different conclusions. However, based on some of the ideas in previous chapters of this book, I would like to offer at least an attempt at a reasonable compromise.

There is no question that legalized abortions avoid the tragedy and butchery of illegal and incompetent "back-alley" abortions, and that in a civilization whose very continuance is threatened by the specter of uncontrolled population growth, widely available medical abortions can serve an important social need. But infanticide would solve both problems and has been employed widely by many human communities, including segments of the classical Greek civilization, which is so generally considered the cultural antecedent of our own. And it is widely practiced today: there are many parts of the world where one out of every four newborn babies does not survive the first year of life. Yet by our laws and mores, infanticide is murder beyond any question. Since a baby born prematurely in the seventh month of pregnancy is in no significant respect different from a fetus *in utero* in the seventh month, it must, it seems to me, follow that abortion, at least in the last trimester, is very close to murder. Objections that the fetus in the third trimester is still not breathing seem specious: Is it permissible to commit infanticide after birth if the umbilicus has not yet been severed, or if the baby has not yet taken its first breath?

Likewise, if I am psychologically unprepared to live with a stranger—in army boot camp or college dormitory, for example—I do not thereby have a right to kill him, and my annoyance at some of the uses of my tax money does not extend to exterminating the recipients of those taxes. The civil liberties point of view is often muddled in such debates. Why, it is sometimes asked, should the beliefs of others on this issue have to extend to me? But those who do not personally support the conventional prohibition against murder are nevertheless required by our society to abide by the criminal code.

On the opposite side of the discussion, the phrase "right to life" is an excellent example of a "buzz word," designed to inflame rather than illuminate. There is no right to life in any society on Earth today, nor has there been at any former time (with a few rare exceptions, such as among the Jains of India). We raise farm animals for slaughter; destroy forests; pollute rivers and lakes until no fish can live there; hunt deer and elk for sport, leopards for their pelts, and whales for dog food; entwine dolphins, gasping and writhing, in great tuna nets; and club seal pups to death for "population management." All these beasts and vegetables are as alive as we. What is protected in many human societies is not life, but human life. And even with this protection, we wage "modern" wars on civilian populations with a toll so terrible we are, most of us, afraid to consider it very deeply. Often such mass murders are justified by racial or nationalistic redefinitions of our opponents as less than human.

In the same way, the argument about the "potential" to be human seems to me particularly weak. Any human egg or sperm under appropriate circumstances has the potential to become a human being. Yet male masturbation and noctural emissions are generally considered natural acts and not cause for murder indictments. In a single ejaculation there are enough spermatozoa for the generation of hundreds of millions of human beings. In addition, it is possible that in the not-too-distant future we may be able to clone a whole human being from a single cell taken from essentially anywhere in the donor's body. If so, any cell in my body has the potential to become a human being if properly preserved until the time of a practical cloning technology. Am I committing mass murder if I prick my finger and lose a drop of blood?

The issues are clearly complex. The solution, equally clearly, must involve a compromise among a number of cherished but conflicting values. The key practical question is to determine when a fetus becomes human. This in turn rests on what we mean by human. Surely not having a human shape, because an artifact of organic materials that resembled a human being but was constructed for the purpose would certainly not be considered human. Likewise, an extraterrestrial intelligent being who did not resemble human beings but who had ethical, intellectual and artistic accomplishments exceeding our own should certainly fall within our prohibitions against murder. It is not what we look like that specifies our humanity, but what we are. The reason we prohibit the killing of human beings must be because of some quality human beings possess, a quality we especially prize, that few or no other organisms on Earth enjoy. It cannot be the ability to feel pain or deep emotions, because that surely extends to many of the animals we gratuitously slaughter.

This essential human quality, I believe, can only be our intelligence. If so, the particular sanctity of human life can be identified with the development and functioning of the neocortex. We cannot require its full development, because that does not occur until many years after birth. But perhaps we might set the transition to humanity at the time when neocortical activity begins, as determined by electroencephalography of the fetus. Some insights on when the brain develops a distinctly human character emerge from the simplest embryological observations (see the figure on page 186). Very little work has been done in this field to date, and it seems to me that such investigations could play a major role in achieving an acceptable compromise in the abortion debate. Undoubt-

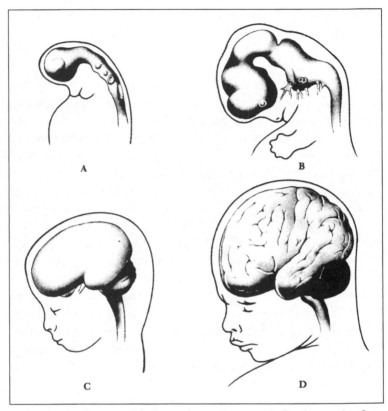

Embryonic development of the human brain. Shown are A after three weeks of gestation; B after seven weeks; C after four months; and D in a newborn infant. The brains in A and B have strong resemblances to the brains of fish and amphibians.

edly there would be a variation from fetus to fetus as to the time of initiation on the first neocortical EEG signals, and a legal definition of the beginning of characteristically human life should be biased conservatively—that is, toward the youngest fetus that exhibits such activity. Perhaps the transition would fall toward the end of the first trimester or near the beginning of the second trimester of pregnancy. (Here we are talking about what, in a rational society, should be prohibited by law: anyone who feels that abortion of a younger fetus might be murder should be under no legal obligation to perform or accept such an abortion.)

But a consistent application of these ideas must avoid human chauvinism. If there are other organisms that share the intelligence of a somewhat backward but fully developed human being, they at least should be offered the same protection against murder that we are willing to extend to human beings late in their uterine existence. Since the evidence for intelligence in dolphins, whales and apes is now at least moderately compelling, any consistent moral posture on abortion should, I would think, include firm strictures against at least the gratuitous slaughter of these animals. But the ultimate key to the solution of the abortion debate would seem to be the investigation of prepartum neocortical activity.

And what of the future evolution of the human brain? There is a wide and growing body of evidence that many forms of mental illness are the result of chemical or wiring malfunctions in the brain. Since many mental diseases have the same symptoms, they may arise from the same malfunctions and should be accessible to the same cures.

The pioneering nineteenth-century British neurologist Hughlings Jackson remarked, "Find out about dreams and you will find out about insanity." Severely dream deprived subjects often begin hallucinating in daytime. Schizophrenia is often accompanied by nighttime sleep impairment, but whether as a cause or an effect is uncertain. One of the most striking aspects of schizophrenia is how unhappy and despairing its sufferers generally are. Might schizophrenia be what happens when the dragons are no longer safely chained at night; when they break the left-hemisphere shackles and burst forth in daylight? Other diseases perhaps result from an impairment of right-hemisphere function: Obsessive-compulsives, for example, are very rarely found to make intuitive leaps.

In the middle 1960s Lester Grinspoon and his colleagues at Harvard Medical School performed a set of controlled experiments on the relative value of various therapeutic techniques for treating schizophrenia. They are psychiatrists, and if they had any bias it was toward the use of verbal rather than pharmacological techniques. But they found to their surprise that the recently developed tranquilizer, thioridazine (one of a group of approximately equally effective antipsychotic drugs known as phenothiazines), was far more effective in controlling if not curing the disease; in fact, they found that thioridazine alone was at least as effective—in the judgment of the patients, their relatives, and the psychiatrists—as thioridazine plus psychotherapy. The integrity of the experimenters in the face of this unexpected finding is breathtaking. (It is difficult to imagine any experiment that would convince leading practitioners of many political or religious philosophies of the superiority of a competing doctrine.)

Recent research shows that endorphins, small protein molecules which occur naturally in the brains of rats and other mammals, can induce in these animals marked muscular rigidity and stupor reminiscent of schizophrenic catatonia. The molecular or neurological cause of schizophrenia—which was once responsible for one out of ten hospital-bed occupancies in the United States—is still unknown; but it is not implausible that someday we will discover precisely what locale or set of neurochemicals in the brain determines this malfunction.

A curious question in medical ethics emerges from the experiments of Grinspoon et al. The tranquilizers are now so effective in treating schizophrenia that it is widely considered unethical to withhold them from a patient. The implication is that the experiments showing tranquilizers to be effective cannot be repeated. It is thought to be an unnecessary cruelty to deny the patient the most successful treatment for his condition. Consequently, there can no longer be a control group of schizophrenics that is not given tranquilizers. If critical experiments in the chemotherapy of brain malfunction can be performed only once, they must be performed the first time very well indeed.

An even more striking example of such chemotherapy is the use of lithium carbonate in the treatment of manic depressives. The ingestion of carefully controlled doses of lithium, the lightest and simplest metal, produces startling improvements—again as reported from the patients' perspective and from the perspective of others—in this agonizing disease. Why so simple a therapy is so strikingly effective is unknown, but it most likely relates to the enzyme chemistry of the brain.

A very strange mental illness is Gilles de la Tourette's disease (named, as always, after the physician who first drew attention to it, not after the most celebrated sufferer of the malady). One of the many motor and speech disorders that are among the symptoms of this disease is a remarkable compulsion to utter—in whatever language the patient is most fluent—an uninterrupted stream of obscenities and profanities. Physicians describe the identification of this disease as "corridor diagnosis": The patient can, with great difficulty, control his compulsion for the length of a short medical visit; as soon as the physician leaves the room for the corridor, the scatologies overflow like the flood from a burst dam. There is a place in the brain that makes "dirty" words (and apes may have it).

There are very few words that the right hemisphere can deal with competently—not much more than hello, goodbye, and … a few choice obscenities. Perhaps Tourette's disease affects the left hemisphere only. The British anthropologist Bernard Campbell of Cambridge University suggests that the limbic system is rather well integrated with the right cerebral hemisphere, which, as we have seen, deals much better with emotions than the left hemisphere does. Whatever else they involve, obscenities carry with them strong emotions. Yet Gilles de la Tourette's disease, complex as it is, seems to be a specific deficiency in a neuronal transmitter chemical, and appears to be alleviated by carefully controlled doses of haloperidol.

Recent evidence indicates that such limbic hormones as ACTH and vasopressin can greatly improve the ability of animals to retain and recall memories. These and similar examples suggest, if not the ultimate perfectability of the brain, at least prospects for its substantial improvement—perhaps through altering the abundance or controlling the production of small brain proteins. Such examples also greatly relieve the burden of guilt commonly experienced by sufferers from a mental disease, a burden rarely felt in victims of, say, measles.

The remarkable fissurization, convolutions and cortical folding of the brain, as well as the fact that the brain fits so snugly into the skull, are clear indications that packing more brain into the present braincase is going to be difficult. Larger brains with larger skulls could not develop until very recently because of limits on the size of the pelvis and the birth canal. But the advent of Caesarean section—performed rarely two thousand years ago but much more commonly today—does permit larger brain volumes. Another possibility is a medical technology sufficiently advanced to permit full-term development of the fetus outside of the uterus. However, the rate of evolutionary change is so slow that none of the problems facing us today is likely to be overcome by significantly larger neocortices and consequent superior intelligences. Before such a time, but not in the immediate future, it may be possible, by brain surgery, to improve those components of the brain we consider worth improving and to inhibit further those components that may be responsible for some of the perils and contradictions facing mankind. But the complexity and redundancy of brain function make such a course of action impractical for the near future, even if it were socially desirable. We may be able to engineer genes before we are able to engineer brains.

It is sometimes suggested that such experiments may provide unscrupulous governments—and there are many of them—with tools to control their citizenry still further. For example, we can imagine a government that implants hundreds of tiny electrodes in the "pleasure" and "pain" centers of the brains of newborn children, electrodes capable of remote radio stimulation—perhaps at frequencies or with access codes known only to the government. When the child grows up, the government might stimulate his pleasure centers if he has performed, in work quota and ideology, an acceptable day's work; otherwise it might stimulate his pain centers. This is a nightmarish vision, but I do not think it is an argument against experiments on electrical stimulation of the brain. It is, rather, an argument against letting the government control the hospitals. Any people that will permit its government to implant such electrodes has already lost the battle and may well deserve what it gets. As in all such technological nightmares, the principal task is to foresee what is possible; to educate the public in its use and misuse; and to prevent its organizational, bureaucratic and governmental abuse.

There is already a range of psychotropic and mood-altering drugs which are, to varying degrees, dangerous or benign (ethyl alcohol is the most widely used and one of the most dangerous), and which appear to act on specific areas of the R-complex, limbic system and neocortex. If present trends continue, even without the encouragement of governments people will pursue the home-laboratory synthesis of and self-experimentation with such drugs—an activity that represents a small further step in our knowledge of the brain, its disorders and untapped potentials.

There is reason to think that many alkaloids and other drugs which affect behavior work by being chemically similar to natural small brain proteins, of which the endorphins are one example. Many of these small proteins act on the limbic system and are concerned with our emotional states. It is now possible to manufacture small proteins made of any specified sequence of amino acids. Thus, the time may soon come when a great variety of molecules will be synthesized capable of inducing human emotional states, including extremely rare ones. For example, there is some evidence that atropine—one of the chief active ingredients in hemlock, foxglove, deadly nightshade, and jimson weed—induces the illusion of flying; and indeed such plants seem to have been the principal constituents of unguents self-administered to the genital mucosa by witches in the Middle Ages—who, rather than actually flying as they boasted, were in fact atropine-tripping. But a vivid hallucination of flying is an extremely specific sensation to be conveyed by a relatively simple molecule. Perhaps there are a range of small proteins which will be synthesized and which will produce emotional states of a sort never before experienced by human beings. This is one of many potential near-term developments in brain chemistry which hold great promise both for good and for evil, depending on the wisdom of those who conduct, control and apply this research.

When I leave my office and get into my car, I find that, unless I make a specific effort of will, I will drive myself home. When I leave home and get into my car, unless I make a similar conscious effort, there is a part of my brain that arranges events so that I end up at my office. If I change my home or my office, after a short period of learning, the new locales supplant the old ones, and whatever brain mechanism controls such behavior has readily adapted to the new coordinates. This is very much like self-programming a part of the brain that works like a digital computer. The comparison is even more striking when we realize that epileptics, suffering from a psychomotor seizure, often go through an exactly comparable set of activities, the only difference being perhaps that they run a few more red lights than I usually do, but have no conscious memory of having performed these actions once the seizure has subsided. Such automatism is a typical symptom of temporal-lobe epilepsy; it also characterizes my first half-hour after awakening. Certainly not all of the brain works like a simple digital computer; the part that *does* the reprogramming, for example, is rather different. But there are enough similarities to suggest that a compatible working arrangement between electronic computers and at least some components of the brain—in an intimate neurophysiological association—can be constructively organized.

The Spanish neurophysiologist José Delgado has devised working feedback loops between electrodes implanted in the brains of chimpanzees and remote electronic computers. Communication between brain and computer is accomplished through a radio link. Miniaturization of electronic computers has now reached the stage where such feedback loops can be "hardwired" and do not require a radio link with a remote computer terminal. For example, it is entirely possible to devise a self-contained feedback loop in which the signs of an oncoming epileptic seizure are recognized and appropriate brain centers are automatically stimulated to forestall or ameliorate the attack. We are not yet at the stage where this is a reliable procedure, but the time when it will be does not seem very far off.

Perhaps some day it will be possible to add a variety of cognitive and intellectual prosthetic devices to the brain—a kind of eyeglasses for the mind. This would be in the spirit of the past accretionary evolution of the brain and is probably far more feasible than attempting to restructure the existing brain. Perhaps one day we will have surgically implanted in our brains small replaceable computer modules or radio terminals which will provide us with a rapid and fluent knowledge of Basque, Urdu, Amharic, Ainu, Albanian, Nu, Hopi, !Kung, or delphinese; or numerical values of the incomplete gamma function and the Tschebysheff polynomials; or the natural history of animal spoor; or all legal precedents for the ownership of floating islands; or radio telepathy connecting several human beings, at least temporarily, in a form of symbiotic association previously unknown to our species.

But the real extensions of our brains, particularly for the uniquely human aspects of the neocortex, are already in the course of being accomplished. Some of them are so old we have forgotten that they have occurred. Rich and unrepressive learning environments for children represent a remarkably promising and successful educational tool. Written language is a notable invention that is essentially a simple machine for the storage and retrieval of quite complex information. The amount of information stored in a large library far exceeds the amount of information in either the human genome or the human brain. The information is certainly not stored as efficiently as it is in biological systems, but it is still serviceably compact, and the development of microfilm, microfiche and the like has greatly improved the extrasomatic information storage capabilities of mankind. The number of bits of information contained in human libraries, works of art, and other cultural institutions would provide a point in the chart on page 22 many factors of 10 beyond the right-hand edge of that figure.

Writing is, however, *very* simple machinery. Much more sophisticated and far more promising extrasomatic information storage and processing lies in the electronic computer. To give some idea of the level of development of such computers and their programs, they can today play perfect tic-tac-toe, world-class checkers, expert chess and practice passable psychotherapy. A computer system now exists which will transcribe a live performance of a classical or other melody into conventional musical notation. The amount of information storage and the speed of information processing by computers are, of course, prodigious.

A non-directive psychotherapy program developed by Joseph Weizenbaum of the Massachusetts Institute of Technology is a good example of the present serviceability of psychiatric computer programs (which are, incidentally, far from the most difficult or most elaborate programs in existence). After the psychotherapy program was developed, a human psychiatrist was brought in to relate in lower-case type (via a typewriter keyboard on an interactive console) with the program, which prints out in upper case. The exchange was probably intended as heuristic but may just possibly have gotten out of hand—on the human side. The conversation begins with the human psychiatrist saying,

doctor, i have terrible news
YOU SAY YOU HAVE TERRIBLE NEWS
yes, there is a computer program at m.i.t. that'll ruin us
WHY DO YOU SAY THAT
i have seen it working
YOU SAY YOU HAVE SEEN IT WORKING
it's frightening—you must see it
YOU THINK I MUST SEE IT
you sound depressed to me
IN YOUR FANTASY PERHAPS WE SOUND DEPRESSED TO EACH OTHER
yes—perhaps
I UNDERSTAND
i do too
YOU SAY YOU DO TOO
we are both in the same boat with respect to that computer
LET'S CONTINUE TO TALK ABOUT YOU—SHALL WE
you are a comfort
WHAT MAKES YOU THINK I AM A COMFORT
at a time like this you want to talk about me
WHY DO YOU THINK I WANT TO TALK ABOUT YOU
because you share my feelings about that damned computer
WHAT DOES THAT SUGGEST TO YOU
that, of course, you hate computers too
YOU THINK I HATE COMPUTERS TOO
why not, they're hardly human, are they
PERHAPS THAT'S WHAT WE ARE HERE TO FIND OUT
no computer could converse as you can, for example
OH
i can't understand how anyone could be fooled by a mere
  machine
CAN'T YOU REALLY
no. after all, there is nothing creative about a computer program
YOU ARE BEING NEGATIVE

i am convinced i am correct
ARE YOU WORRIED TO HAVE BEEN CONVINCED YOU ARE
     CORRECT
no, not in the least. not at all. not even a little.
YOU ARE BEING NEGATIVE
you said you agreed with me, didn't you
IS THAT WHAT YOU HEARD ME SAY
why don't you try to dissuade me—computers diddle with
     numbers, that's all
YOU WISH ME TO TRY TO DISSUADE YOU
you can't
YOU THINK I CAN'T
for example, could a computer in a conversation like this raise
     a new topic
YOU SEEM TO HAVE AVOIDED SPEAKING OF YOUR PARENTS
     ALTOGETHER
precisely as you just did
YOU THINK I JUST DID
yes, and that's what shows the difference between a man and a
     computer
I SEE

The computer psychiatrist lacks sophistication, training, sensitivity, human intuition; it is mechanical (of course!), inflexible
and relatively unresponsive to emotional nuances and nonverbal
cues. And yet it has produced a conversation more intelligent than
many. Its response on "a new topic" is stunning. But that response
is very likely only a fluke. The program is undoubtedly designed
to pay attention to words such as "mother," "father," "parent," and
so on; after the computer's clock has ticked away so many minutes, if these words have not been introduced, the program is
designed to come up with "You seem to have avoided ..." Emerging at just the moment it did, the remark gives an eerie impression
of insight.

But what is the game of psychotherapy if not a very complex, learned set of responses to human situations? Is not the psychiatrist also preprogrammed to give certain responses? Non-directive psychotherapy clearly requires very simple computer programs, and the appearance of insight requires only slightly more sophisticated programs. I do not intend these remarks to disparage the psychiatric profession in any way, but rather to augur the coming of machine intelligence. Computers are by no means yet at a high enough level of development to recommend the widespread use of computer psychotherapy. But it does not seem to me a forlorn hope that we may one day have extremely patient, widely available and, at least for certain problems, adequately competent computer therapists. Some programs already in existence are given high marks by patients because the therapist is perceived as unbiased and extremely generous with his or her or its time.

Computers are now being developed in the United States that will be able to detect and diagnose their own malfunctions. When systematic performance errors are found, the faulty components will be automatically bypassed or replaced. Internal consistency will be tested by repeated operation and through standard programs whose consequences are known independently; repair will be accomplished chiefly by redundant components. There are already in existence programs—e.g., in chess-playing computers—capable of learning from experience and from other computers. As time goes on, the computer appears to become increasingly intelligent. Once the programs are so complex that their inventors cannot quickly predict all possible responses, the machines will have the appearance of, if not intelligence, at least free will. Even the computer on the Viking Mars lander, which has a memory of only 18,000 words, is at this point of complexity: we do not in all cases know what the computer will do with a given command. If we knew, we would say it is "only" or "merely" a computer. When we do not know, we begin to wonder if it is truly intelligent.

The situation is very much like the commentary that has echoed over the centuries after a famous animal story told both by Plutarch and by Pliny: A dog, following the scent of its master, was observed to come to a triple fork in the road. It ran down the leftmost prong, sniffing; then stopped and returned to follow the middle prong for a short distance, again sniffing and then turning back. Finally, with no sniffing at all, it raced joyously down the right-hand prong of the forked road.

Montaigne, commenting on this story, argued that it showed clear canine syllogistic reasoning: My master has gone down one of these roads. It is not the left-hand road; it is not the middle road; therefore it must be the right-hand road. There is no need for me to corroborate this conclusion by smell—the conclusion follows by straightforward logic.

The possibility that reasoning at all like this might exist in the animals, although perhaps less clearly articulated, was troubling to many, and long before Montaigne, St. Thomas Aquinas attempted unsuccessfully to deal with the story. He cited it as a cautionary example of how the appearance of intelligence can exist where no intelligence is in fact present. Aquinas did not, however, offer a satisfactory alternative explanation of the dog's behavior. In human split-brain patients, it is quite clear that fairly elaborate logical analysis can proceed surrounded by verbal incompetence.

We are at a similar point in the consideration of machine intelligence. Machines are just passing over an important threshold: the threshold at which, to some extent at least, they give an unbiased human being the impression of intelligence. Because of a kind of human chauvinism or anthropocentrism, many humans are reluctant to admit this possibility. But I think it is inevitable. To me it is not in the least demeaning that consciousness and intelligence are the result of "mere" matter sufficiently complexly arranged; on the contrary, it is an exalting tribute to the subtlety of matter and the laws of Nature.

It by no means follows that computers will in the immediate future exhibit human creativity, subtlety, sensitivity or wisdom. A classic and probably apocryphal illustration is in the field of machine translation of human languages: a language—say, English—is input and the text is output in another language—say, Chinese. After the completion of an advanced translation program, so the story goes, a delegation which included a U.S. senator was proudly taken through a demonstration of the computer system. The senator was asked to produce an English phrase for translation and promptly suggested, "Out of sight, out of mind." The machine dutifully whirred and winked and generated a piece of paper on which were printed a few Chinese characters. But the senator could not read Chinese. So, to complete the test, the program was run in reverse, the Chinese characters input and an English phrase output. The visitors crowded around the new piece of paper, which to their initial puzzlement read: "Invisible idiot."

Existing programs are only marginally competent even on matters of this not very high degree of subtlety. It would be folly to entrust major decisions to computers at our present level of development—not because the computers are not intelligent to a degree, but because, in the case of most complex problems, they will not have been given all relevant information. The reliance on computers in determining American policy and military actions during the Vietnam war is an excellent example of the flagrant misuse of these machines. But in reasonably restricted contexts the human use of artificial intelligence seems to be one of the two practicable major advances in human intelligence available in the near future. (The other is enrichment of the preschool and school learning environments of children.)

Those who have not grown up with computers generally find them more frightening than those who have. The legendary manic computer biller who will not take no—or even yes—for an answer, and who can be satisfied only by receiving a check for zero dollars and zero cents is not to be considered representative of the entire tribe; it is a feeble-minded computer to begin with, and its mistakes are those of its human programmers. The growing use in North America of integrated circuits and small computers for aircraft safety, teaching machines, cardiac pacemakers, electronic games, smoke-actuated fire alarms and automated factories, to name only a few uses, has helped greatly to reduce the sense of strangeness with which so novel an invention is usually invested. There are some 200,000 digital computers in the world today; in another decade, there are likely to be tens of millions. In another generation, I think that computers will be treated as a perfectly natural—or at least commonplace—aspect of our lives.

Consider, for example, the development of small, pocket computers. I have in my laboratory a desk-sized computer purchased with a research grant in the late 1960s for $4,900. I also have another product of the same manufacturer, a computer that fits into the palm of my hand, which was purchased in 1975. The new computer does everything that the old computer did, including programming capability and several addressable memories. But it costs $145, and is getting cheaper at a breathtaking rate. That represents quite a spectacular advance, both in miniaturization and in cost reduction, in a period of six or seven years. In fact, the present limit on the size of hand-held computers is the requirement that the buttons be large enough for our somewhat gross and clumsy human fingers to press. Otherwise, such computers could easily be built no larger than my fingernail. Indeed, ENIAC, the first large electronic digital computer, constructed in 1946, contained 18,000 vacuum tubes and occupied a large room. The same computational ability resides today in a silicon chip microcomputer the size of the smallest joint of my little finger.

The speed of transmission of information in the circuitry of such computers is the velocity of light. Human neural transmission is one million times slower. That in nonarithmetic operations the small and slow human brain can still do so much better than the large and fast electronic computer is an impressive tribute to how cleverly the brain is packaged and programmed—features brought about, of course, by natural selection. Those who possessed poorly programmed brains eventually did not live long enough to reproduce.

Computer graphics have now reached a state of sophistication that permits important and novel kinds of learning experiences in arts and sciences, and in both cerebral hemispheres. There are individuals, many of them analytically extremely gifted, who are impoverished in their abilities to perceive and imagine spatial relations, particularly three-dimensional geometry. We now have computer programs that can gradually build up complex geometrical forms before our eyes and rotate them on a television screen connected to the computer.

At Cornell University, such a system has been designed by Donald Greenberg of the School of Architecture. With this system it is possible to draw a set of regularly spaced lines which the computer interprets as contour intervals. Then, by touching our light pen to any of a number of possible instructions on the screen, we command the construction of elaborate three-dimensional images which can be made larger or smaller, stretched in a given direction, rotated, joined to other objects or have designated parts excised. (See figures on pp. 202–203.) This is an extraordinary tool for improving our ability to visualize three-dimensional forms—a skill extremely useful in graphic arts, in science and in technology. It also represents an excellent example of cooperation between the two cerebral hemispheres: the computer, which is a supreme construction of the left hemisphere, teaches us pattern recognition, which is a characteristic function of the right hemisphere.

There are other computer programs that exhibit two- and three-dimensional projections of four-dimensional objects. As the four-dimensional objects turn, or our perspective changes, not only do we see new parts of the four-dimensional objects; we also seem to see the synthesis and destruction of entire

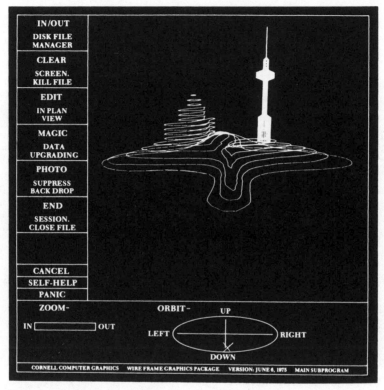

Example of a simple computer graphics routine. Each figure was created solely by drawing free-hand contours with a "light pen" on a television screen. The computer converted this into perspective drawings in elevation from any view angle—directly from the side of this free-form sculpture at left and at an angle at right. The tower

geometrical subunits. The effect is eerie and instructive and helps to make four-dimensional geometry much less mysterious; we are not nearly so baffled as I imagine a mythical two-dimensional creature would be on encountering the typical projection (two squares with the corners connected) of a three-dimensional cube on a flat surface. The classical artistic problem of

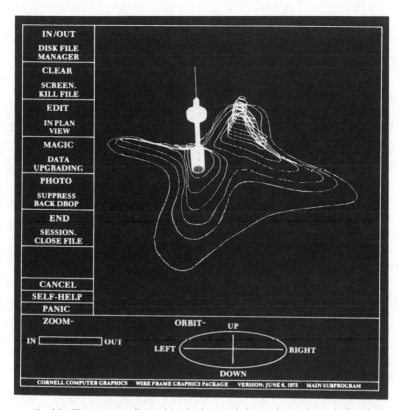

was "webbed" automatically, and is tilted toward the reader in the right-hand diagram. In addition to a full capability for rotation and zoom, the observer can request with his "light pen" orthogonal, perspective, or stereoscopic dynamic images (Program WIRE by Mare Levoy, Laboratory of Computer Graphics, Cornell University).

perspective—the projection of three-dimensional objects onto two-dimensional canvases—is enormously clarified by computer graphics; the computer is obviously also a major tool in the quite practical problem of picturing an architect's design of a building, made in two dimensions, from all vantage points in three dimensions.

Computer graphics are now being extended into the area of play. There is a popular game, sometimes called Pong, which simulates on a television screen a perfectly elastic ball bouncing between two surfaces. Each player is given a dial that permits him to intercept the ball with a movable "racket." Points are scored if the motion of the ball is not intercepted by the racket. The game is very interesting. There is a clear learning experience involved which depends exclusively on Newton's second law for linear motion. As a result of Pong, the player can gain a deep intuitive understanding of the simplest Newtonian physics—a better understanding even than that provided by billiards, where the collisions are far from perfectly elastic and where the spinning of the pool balls interposes more complicated physics.

This sort of information gathering is precisely what we call play. And the important function of play is thus revealed: it permits us to gain, without any particular future application in mind, a holistic understanding of the world, which is both a complement of and a preparation for later analytical activities. But computers permit play in environments otherwise totally inaccessible to the average student.

A still more interesting example is provided by the game Space War, whose development and delights have been chronicled by Stuart Brand. In Space War, each side controls one or more "space vehicles" which can fire missiles at the other. The motions of both the spacecraft and the missiles are governed by certain rules—for example, an inverse square gravitational field set up by a nearby "planet." To destroy the spaceship of your opponent you must develop an understanding of Newtonian gravitation that is simultaneously intuitive and concrete. Those of us who do not frequently engage in interplanetary space flight do not readily evolve a right-hemisphere comprehension of Newtonian gravitation. Space War can fill that gap.

The two games, Pong and Space War, suggest a gradual elaboration of computer graphics so that we gain an experiential and intuitive understanding of the laws of physics. The laws of physics are almost always stated in analytical and algebraic—that is to say, left-hemisphere—terms; for example, Newton's second law is written $F = m\,a$, and the inverse square law of gravitation as $F = G\,M\,m/r^2$. These analytical representations are extremely useful, and it is certainly interesting that the universe is made in such a way that the motion of objects can be described by such relatively simple laws. But these laws are nothing more than abstractions from experience. Fundamentally they are mnemonic devices. They permit us to remember in a simple way a great range of cases that would individually be much more difficult to remember—at least in the sense of memory as understood by the left hemisphere. Computer graphics gives the prospective physical or biological scientist a wide range of experience with the cases his laws of nature summarize; but its most important function may be to permit those who are not scientists to grasp in an intuitive but nevertheless deep manner what the laws of nature are about.

There are many non-graphical interactive computer programs which are extremely powerful teaching tools. The programs can be devised by first-rate teachers, and the student has, in a curious sense, a much more personal, one-to-one relationship with the teacher than in the usual classroom setting; he may also be as slow as he wishes without fear of embarrassment. Dartmouth College employs computer learning techniques in a very broad array of courses. For example, a student can gain a deep insight into the statistics of Mendelian genetics in an hour with the computer rather than spend a year crossing fruit flies in the laboratory. Another student can examine the statistical likelihood of becoming pregnant were she to use various birth control methods. (This program has built into it a one-in-ten-billion chance of a woman's becoming pregnant when strictly celibate, to allow for contingencies beyond present medical knowledge.)

The computer terminal is a commonplace on the Dartmouth campus. A very high proportion of Dartmouth undergraduates learn not only to use such programs but also to write their own. Interaction with computers is widely viewed as more like fun than like work, and many colleges and universities are in the process of imitating and extending Dartmouth's practice. Dartmouth's preeminence in this innovation is related to the fact that its president, John G. Kemeny, is a distinguished computer scientist and the inventor of a very simple computer language called BASIC.

The Lawrence Hall of Science is a kind of museum connected with the University of California at Berkeley. In its basement is a rather modest room filled with about a dozen inexpensive computer terminals, each hooked up to a time-sharing mini-computer system located elsewhere in the building. Reservations for access to these terminals are sold for a modest fee, and they may be made up to one hour in advance. The clientele is predominantly youngsters, and the youngest are surely less than ten years old. A very simple interactive program available there is the game Hangman. To play Hangman you type on a fairly ordinary typewriter keyboard the computer code "XEQ-$HANG". The computer then types out:

HANGMAN
CARE FOR THE RULES?

If you type "YES", the machine replies:

GUESS A LETTER IN THE WORD I'M THINKING OF.
IF YOU ARE RIGHT, THEN I WILL TELL YOU. BUT
IF YOU ARE WRONG (HA, HA) YOU WILL BE CLOSER
(SNICKER, SNICKER) TO DEATH BY HANGING!
THE WORD HAS EIGHT LETTERS.
YOUR GUESS IS ...?

Let us say you type the response: "E". The computer then types:

— — — — — — — E

If you guess wrong, the computer then types out an engaging simulacrum (within the limitations of the characters available to it) of a human head. And in the usual manner of the game there is a race between the gradually emerging word and the gradually emerging form of a human being about to be hanged.

In two games of Hangman I recently witnessed, the correct answers were "VARIABLE" and "THOUGHT". If you win the game the program—true to its mustache-twirling villainy—types out a string of non-letter characters from the top row of the typewriter keyboard (used in comic books to indicate curses) and then prints:

RATS, YOU WIN
CARE FOR ANOTHER CHANCE TO DIE?

Other programs are more polite. For example, "XEQ-$KING" yields:

THIS IS THE ANCIENT KINGDOM OF SUMERIA, AND YOU
ARE ITS VENERATED RULER. THE FATE OF SUMERIA'S
ECONOMY AND OF YOUR LOYAL SUBJECTS IS ENTIRELY
IN YOUR HANDS. YOUR MINISTER, HAMMURABI, WILL
REPORT TO YOU EACH YEAR ON POPULATION AND
ECONOMY. USING HIS INFORMATION YOU MUST LEARN
TO ALLOCATE RESOURCES FOR YOUR KINGDOM WISELY.
SOMEONE IS ENTERING YOUR COUNCIL CHAMBER ...

Hammurabi then presents you with relevant statistics on the number of acres owned by the city, how many bushels per acre were harvested last year, how many were destroyed by rats, how many are now in storage, what the present population is, how many people died of starvation last year, and how many migrated to the city. He begs to inform you of the current exchange rate of land for food and queries how many acres you wish to buy. If you ask for too much, the program prints:

A statue of Gudea, the Neo-Sumerian governor of Lagash, about 2150 B.C. Cuneiform writing, which covers Gudea's robe, was widespread in this era, the Third Dynasty of Ur, a time of maritime trade, commercial exuberance, and the earliest known legal code—all intimately connected with the proliferation of literacy.

HAMMURABI: PLEASE THINK AGAIN. YOU HAVE ONLY
TWENTY-EIGHT HUNDRED BUSHELS IN STORE.

Hammurabi turns out to be an extremely patient and polite Grand Vizier. As the years flicker by, you gain a powerful impression that it may be very difficult, at least in certain market economies, to increase both the population and landholdings of a state while avoiding poverty and starvation.

Among the many other programs available is one called Grand Prix Racing which permits you to choose from among a range of opponents, running from a Model T Ford to a 1973 Ferrari. If your speed or acceleration are too low at appropriate places on the track, you lose; if too high, you crash. Since distances, velocities and accelerations must be given explicitly, there is no way to play this game without learning some physics. The array of possible courses of computer interactive learning is limited only by the ingenuity of the programmers, and that is a well that runs very deep.

Since our society is so profoundly influenced by science and technology, which the bulk of our citizens understand poorly or not at all, the widespread availability in both schools and homes of inexpensive interactive computer facilities could just possibly play an important role in the continuance of our civilization.

The only objection I have ever heard to the widespread use of pocket calculators and small computers is that, if introduced to children too early, they preempt the learning of arithmetic, trigonometry and other mathematical tasks that the machine is able to perform faster and more accurately than the student. This debate has occurred before.

In Plato's *Phaedrus*—the same Socratic dialogue I referred to earlier for its metaphor of chariot, charioteer and two horses—there is a lovely myth about the god Thoth, the Egyptian equivalent of Prometheus. In the tongue of ancient Egypt, the phrase that

Example of early Egyptian hieroglyphics from a tablet of Sesostris I at Karnak.

designates written language means literally "The Speech of the Gods." Thoth is discussing his invention* of writing with Thamus (also called Ammon), a god-king who rebukes him in these words:

> This discovery of yours will create forgetfulness in the learn-
> ers' souls, because they will not use their memories; they will

---

*According to the Roman historian Tacitus, the Egyptians claimed to have taught the alphabet to the Phoenicians, "who, controlling the seas, introduced it to Greece and were credited with inventing what they had really borrowed." According to legend, the alphabet arrived in Greece with Cadmus, Prince of Tyre, seeking his sister, Europa, who had been stolen away to the island of Crete by Zeus, king of the gods, temporarily disguised as a bull. To protect Europa from those who would steal her back to Phoenicia, Zeus ordered a bronze robot made which, with clanking steps, patrolled Crete and turned back or sank all approaching foreign vessels. Cadmus, however, was elsewhere—unsuccessfully seeking his sister in Greece when a dragon devoured all his men; whereupon he slew the dragon and, in response to instructions from the goddess Athena, sowed the dragon's teeth in the furrows of a plowed field. Each tooth became a warrior; and Cadmus and his men together founded Thebes, the first civilized Greek city, bearing the same name as one of the two capital cities of ancient Egypt. It is curious to find in the same legendary account the invention of writing, the founding of Greek civilization, the first known reference to artificial intelligence, and the continuing warfare between humans and dragons.

trust to the external written characters and not remember of
themselves. The specific which you have discovered is an aid
not to memory, but to reminiscence, and you give your dis-
ciples not truth, but only the semblance of truth; they will
be hearers of many things and will have learned nothing;

A microprocessing unit of a microcomputer, about half a centimeter on a side. It
is an integrated circuit deposited on a single crystal silicon chip and containing
about 5,400 transistors.

they will appear to be omniscient and will generally know nothing; they will be tiresome company, having the show of wisdom without its reality.

I am sure there is some truth to Thamus' complaint. In our modern world, illiterates have a different sense of direction, a different sense of self-reliance, and a different sense of reality. But before the invention of writing, human knowledge was restricted to what one person or a small group could remember. Occasionally, as with the Vedas and the two great epic poems of Homer, a substantial body of information could be preserved. But there were, so far as we know, few Homers. After the invention of writing, it was possible to collect, integrate and utilize the accumulated wisdom of all times and peoples; humans were no longer restricted to what they and their immediate acquaintances could remember. Literacy gives us access to the greatest and most influential minds in history: Socrates, say, or Newton have had audiences vastly larger than the total number of people either met in his whole lifetime. The repeated rendering of an oral tradition over many generations inevitably leads to errors in transmission and the gradual loss of the original content, a degradation of information that occurs far more slowly with the successive reprinting of written accounts.

Books are readily stored. We can read them at our own pace without disturbing others. We can go back to the hard parts, or delight once again in the particularly enjoyable parts. They are mass-produced at relatively low cost. And reading itself is an amazing activity: You glance at a thin, flat object made from a tree, as you are doing at this moment, and the voice of the author begins to speak inside your head. (Hello!) The improvement in human knowledge and survival potential following the invention of writing was immense. (There was also an improvement in self-reliance: It is possible to learn at least the rudiments of an art or a science from a book and not be dependent on the lucky accident that there is a nearby master craftsman to whom we may apprentice ourselves.)

When all is said and done, the invention of writing must be reckoned not only as a brilliant innovation but as a surpassing good for humanity. And assuming that we survive long enough to use their inventions wisely, I believe the same will be said of the modern Thoths and Prometheuses who are today devising computers and programs at the edge of machine intelligence. The next major structural development in human intelligence is likely to be a partnership between intelligent humans and intelligent machines.

The silent hours steal on ...

WM. SHAKESPEARE
King Richard III

The question of all questions for humanity, the problem which lies behind all others and is more interesting than any of them is that of the determination of man's place in Nature and his relation to the Cosmos. Whence our race came, what sorts of limits are set to our power over Nature and to Nature's power over us, to what goal we are striving, are the problems which present themselves afresh, with undiminished interest, to every human being born on earth.

T. H. HUXLEY, 1863

CHAPTER IX

KNOWLEDGE
IS OUR DESTINY:
TERRESTRIAL AND
EXTRATERRESTRIAL
INTELLIGENCE

AND SO at last I return to one of the questions with which I started: the search for extraterrestrial intelligence. While the suggestion is sometimes made that the preferred channel of interstellar discourse will be telepathic, this seems to me at best a playful notion. At any rate, there is not the faintest evidence in support of it; and I have yet to see even moderately convincing evidence for telepathic transmission on *this* planet. We are not yet capable of significant interstellar space flight, although some other more advanced civilization might be. Despite all the talk of unidentified flying objects and ancient astronauts, there is no serious evidence that we have been or are being visited.

That, then, leaves machines. Communication with extraterrestrial intelligence may employ the electromagnetic spectrum, and most likely the radio part of the spectrum; or it might employ gravity waves, neutrinos, just conceivably tachyons (if they exist), or some new aspect of physics that will not be discovered for another three centuries. But whatever the channel, it will require machines to use, and if our experience in radioastronomy is any guide, computer-actuated machines with abilities approaching what we might call intelligence. To run through many days' worth of data on 1,008 different frequencies, where the information may vary every few seconds or faster, cannot be done well by visually scanning the records. It requires autocorrelation techniques and large electronic computers. And this situation, which applies to observations that Frank Drake of Cornell and I have recently performed at the Arecibo Observatory, can only become more complex—that is, more dependent on computers—with the listening devices likely to be employed in the near future. We can design receiving and transmitting programs of immense complexity. If we are lucky we will employ stratagems of great cleverness and elegance. But we cannot avoid utilizing the remarkable capabilities of machine intelligence if we wish to search for extraterrestrial intelligence.

The number of advanced civilizations in the Milky Way Galaxy today depends on many factors, ranging from the number of planets per star to the likelihood of the origin of life. But once life has started in a relatively benign environment and billions of years of evolutionary time are available, the expectation of many of us is that intelligent beings would develop. The evolutionary path would, of course, be different from that taken on Earth. The precise sequence of events that have taken place here—including the extinction of the dinosaurs and the recession of the Pliocene and Pleistocene forests—have probably not occurred in precisely the same way anywhere else in the entire universe. But there should be many functionally equivalent pathways to a similar end result. The entire evolutionary record on our planet, particularly the record contained in fossil endocasts, illustrates a progressive tendency toward intelligence. There is nothing mysterious about this: smart organisms by and

"Stars" by M. C. Escher.

large survive better and leave more offspring than stupid ones. The details will certainly depend on circumstances, as, for example, if nonhuman primates with language have been exterminated by humans, while slightly less communicative apes ignored by our ancestors. But the general trend seems quite clear and should apply to the evolution of intelligent life elsewhere. Once intelligent beings achieve technology and the capacity for self-destruction of their species, the selective advantage of intelligence becomes more uncertain.

And what if we receive a message? Is there any reason to think that the transmitting beings—evolved over billions of years of geological time in an environment vastly different from our own—would be sufficiently similar to us for their messages to be understood? I think the answer must be yes. A civilization transmitting radio messages must at least know about radio. The frequency, time constant, and bandpass of the message are common to transmitting and receiving civilizations. The situation may be a little like that of amateur or ham radio operators. Except for occasional emergencies, their conversations seem almost exclusively concerned with the mechanics of their instruments: it is the one aspect of their lives they are certain to have in common.

But I think the situation is far more hopeful than this. We know that the laws of nature—or at least many of them—are the same everywhere. We can detect by spectroscopy the same chemical elements, the same common molecules on other planets, stars and galaxies; and the fact that the spectra are the same shows that the same mechanisms by which atoms and molecules are induced to absorb and emit radiation exist everywhere. Distant galaxies can be observed moving ponderously about each other in precise accord with the same laws of gravitation that determine the motion of a tiny artificial satellite about our pale blue planet Earth. Gravity, quantum mechanics, and the great bulk of physics and chemistry are observed to be the same elsewhere as here.

Intelligent organisms evolving on another world may not be like us biochemically. They will almost certainly have evolved significantly different adaptations—from enzymes to organ systems—to deal with the different circumstances of their several worlds. But they must still come to grips with the same laws of nature.

The laws of falling bodies seem simple to us. At constant acceleration, as provided by Earth's gravity, the velocity of a falling object increases proportional to the time; the distance fallen proportional to the square of the time. These are very elementary relations. Since Galileo at least, they have been fairly generally grasped. Yet we can imagine a universe in which the laws of nature are immensely more complex. But we do not live in such a universe. Why not? I think it may be because all those organisms who perceived their universe as very complex are dead. Those of our arboreal ancestors who had difficulty computing their trajectories as they brachiated from tree to tree did not leave many offspring. Natural selection has served as a kind of intellectual sieve, producing brains and intelligences increasingly competent to deal with the laws of nature. This resonance, extracted by natural selection, between our brains and the universe may help explain a quandary set by Einstein: The most incomprehensible property of the universe, he said, is that it is so comprehensible.

If this is so, the same evolutionary winnowing must have occurred on other worlds that have evolved intelligent beings. Extraterrestrial intelligences that lack avian or arboreal ancestors may not share our passion for space flight. But all planetary atmospheres are relatively transparent in the visible and radio parts of the spectrum—because of the quantum mechanics of the cosmically most abundant atoms and molecules. Organisms throughout the universe should therefore be sensitive to optical and/or radio radiation, and, after the development of physics, the idea of electromagnetic radiation for interstellar communication should be a cosmic commonplace—a convergent idea evolving independently on countless worlds throughout the galaxy after the local discovery of elementary astronomy, what we might call the facts of life. If we are fortunate enough to make contact with some of those other beings, I think we will find that much of their biology, psychology, sociology and politics will seem to us stunningly exotic and deeply mysterious. But I suspect we will have little difficulty in understanding each other on the simpler aspects of astronomy, physics, chemistry and perhaps mathematics.

I would certainly not expect their brains to be anatomically or physiologically or perhaps even chemically close to ours. Their brains will have had different evolutionary histories in different environments. We have only to look at terrestrial beasts with substantially different organ systems to see how much variation in brain physiology is possible. There is, for example, an African fresh-water fish, the Mormyrid, which often lives in murky water where visual detection of predators, prey or mates is difficult. The Mormyrid has developed a special organ which establishes an electric field and monitors that field for any creatures traversing it. This fish possesses a cerebellum that covers the entire back of its brain in a thick layer reminiscent of the neocortex of mammals. The Mormyrids have a spectacularly different sort of brain, and yet in the most fundamental biological sense they are far more closely related to us than any intelligent extraterrestrial beings.

The brains of extraterrestrials will probably have several or many components slowly accreted by evolution, as ours have. There may still be a tension among their components as among ours, although the hallmark of a successful, long-lived civilization may be the ability to achieve a lasting peace among the several brain components. They almost certainly will have significantly extended their intelligence extrasomatically, by employing intelligent machines. But I think it highly probable that our brains and machines and their brains and machines will ultimately understand one another very well.

The practical benefits as well as the philosophical insights likely to accrue from the receipt of a long message from an advanced civilization are immense. But how great the benefits and how fast we can assimilate them depend on the details of the message contents, about which it is difficult to make reliable predictions. One consequence, however, seems clear; the receipt of a message from an advanced civilization will show that there *are* advanced civilizations, that there are methods of avoiding the self-destruction that seems so

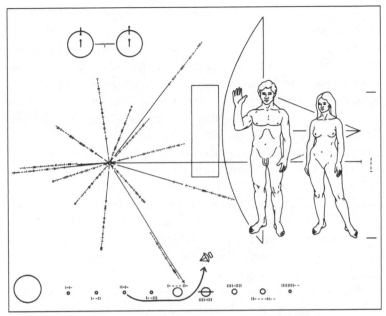

The plaque aboard the Pioneer 10 and 11 spacecraft, the first vehicles of mankind to venture into interstellar space. The 6-by-9-inch gold anodized aluminum plaques convey, in what is hoped is easily understood scientific language, some information on the locale, epoch, and nature of the builders of the spacecraft. Interstellar radio messages can be much richer in information content than this message in a bottle cast into the cosmic ocean.

real a danger of our present technological adolescence. Thus the receipt of an interstellar message would provide a very practical benefit that in mathematics is called the existence theorem—in this case the demonstration that it is possible for societies to live and prosper with advanced technology. Finding a solution to a problem is helped enormously by the certain knowledge that a solution exists. This is one of many curious connections between the existence of intelligent life elsewhere and the existence of intelligent life on Earth.

While more rather than less knowledge and intelligence seems so clearly the only way out of our present difficulties and the only aperture to a significant future for mankind (or indeed to any future at all), this is not a view always adopted in practice. Governments often lose sight of the difference between short-term and long-term benefits. The most important practical benefits have come about from the most unlikely and apparently impractical scientific advances. Radio is today not only the prime channel in the search for extra-terrestrial intelligence, it is the means by which emergencies are responded to, news is transmitted, telephone calls relayed and global entertainment aired. Yet radio came about because a Scottish physicist, James Clerk Maxwell, invented a term, which he called the displacement current, in a set of partial differential equations now known as Maxwell's equations. He proposed the displacement current essentially because the equations were aesthetically more appealing with it than without it.

The universe is intricate and elegant. We wrest secrets from nature by the most unlikely routes. Societies will, of course, wish to exercise prudence in deciding which technologies—that is, which applications of science—are to be pursued and which not. But without funding basic research, without supporting the acquisition of knowledge for its own sake, our options become dangerously limited. Only one physicist in a thousand need stumble upon something like the displacement current to make the support of all thousand a superb investment for society. Without vigorous, farsighted and continuing encouragement of fundamental scientific research, we are in the position of eating our seed corn: we may fend off starvation for one more winter, but we have removed the last hope of surviving the following winter.

In a time in some respects similar to our own, St. Augustine of Hippo, after a lusty and intellectually inventive young manhood, withdrew from the world of sense and intellect and advised others to do likewise: "There is another form of temptation, even more fraught with danger. This is the disease of curiosity.... It is this which drives us on to try to discover the secrets of nature, those secrets which are beyond our understanding, which can avail us nothing and which men should not wish to learn. . . . In this immense forest, full of pit-falls and perils, I have drawn myself back, and pulled myself away from these thorns. In the midst of all these things which float unceasingly around me in everyday life, I am never surprised at any of them, and never captivated by my genuine desire to study them.... I no longer dream of the stars." The time of Augustine's death, 430 A.D., marks the beginning of the Dark Ages in Europe.

In the last chapter of *The Ascent of Man* Bronowski confessed himself saddened "to find myself suddenly surrounded in the West by a sense of terrible loss of nerve, a retreat from knowledge." He was talking, I think, partly about the very limited understanding and appreciation of science and technology—which have shaped our lives and civilizations—in public and political communities; but also about the increasing popularity of various forms of marginal, folk- or pseudo-science, mysticism and magic.

There is today in the West (but not in the East) a resurgent interest in vague, anecdotal and often demonstrably erroneous doctrines that, if true, would betoken at least a more interesting universe, but that, if false, imply an intellectual carelessness, an absence of tough-mindedness, and a diversion of energies not very promising for our survival. Such doctrines include astrology (the view that which stars, one hundred trillion miles away, are rising at the moment of my birth in a closed building affect my destiny profoundly); the Bermuda Triangle "mystery" (which holds in many versions that an unidentified flying object lives in the ocean off Bermuda and eats ships and airplanes); flying saucer accounts in general; the belief in ancient astronauts; the photography of ghosts; pyramidology (including the view that my razor blade stays sharper within a cardboard pyramid than within a cardboard cube); Scientology; auras and Kirlian photography; the emotional lives and musical preferences of geraniums; psychic surgery; flat and hollow earths; modern prophecy; remote cutlery warping; astral projections; Velikovskian catastrophism; Atlantis and

Mu; spiritualism; and the doctrine of the special creation, by God or gods, of mankind despite our deep relatedness, both in biochemistry and in brain physiology, with the other animals. It may be that there are kernels of truth in a few of these doctrines, but their widespread acceptance betokens a lack of intellectual rigor, an absence of skepticism, a need to replace experiments by desires. These are by and large, if I may use the phrase, limbic and right-hemisphere doctrines, dream protocols, natural—the word is certainly perfectly appropriate—and human responses to the complexity of the environment we inhabit. But they are also mystical and occult doctrines, devised in such a way that they are not subject to disproof and characteristically impervious to rational discussion. In contrast, the aperture to a bright future lies almost certainly through the full functioning of the neocortex— reason alloyed with intuition and with limbic and R-complex components, to be sure, but reason nonetheless: a courageous working through of the world as it really is.

It is only in the last day of the Cosmic Calendar that substantial intellectual abilities have evolved on the planet Earth. The coordinated functioning of both cerebral hemispheres is the tool Nature has provided for our survival. We are unlikely to survive if we do not make full and creative use of our human intelligence.

"We are a scientific civilization," declared Jacob Bronowski. "That means a civilization in which knowledge and its integrity are crucial. Science is only a Latin word for knowledge. . . . Knowledge is our destiny."

ACKNOWLEDGMENTS

To write a book on a subject so far from one's primary training is at best incautious. But, as I have tried to explain, the temptation was irresistible. Whatever virtues this book may have are largely thanks to those who performed the fundamental research described, and to those professionals in the biological and social sciences who were kind enough to read and react to my arguments. I am indebted for critical comments and stimulating discussions to the late L. S. B. Leakey and Hans-Lukas Teuber, to Joshua Lederberg, James Maas, John Eisenberg, Bernard Campbell, Lester and David Grinspoon, Stephen Jay Gould, William Dement, Geoffrey Bourne, Philip Morrison, Charles Hockett, Ernest Hartmann, Richard Gregory, Paul Rozin, Jon Lomberg, Timothy Ferris, and particularly to Paul MacLean. I appreciate the painstaking care which several of them, as well as editor Anne Freedgood and copy editor Nancy Inglis, both at Random House, took in reading earlier drafts of this book. They are, it is probably unnecessary to add, not to be held responsible for my speculations or for any errors which may be found herein. I am grateful to Linda Sagan and Sally Forbes for picture research; to several colleagues for preprints of scientific reports in advance of publication; and to Don Davis for the cover painting, which is intended not as a literal depiction of any particular epoch in Earth history, but as a metaphor of a few of the ideas set forth above. Some of this work was made possible by the institution of sabbatic leave at Cornell University. I am also grateful for their kind hospitality to L. E. H. Trainor, M. Silverman, C. Lumsden and Andrew Baines, Principal of New College, all affiliated with the University of Toronto.

Substantial parts of Chapter 1 appeared in the magazine *Natural History*. Some of the ideas in this book were first presented at a joint colloquium of the Massachusetts Mental Health Center and Harvard University Medical School's Department of Psychiatry, and at an L. S. B. Leakey Foundation lecture at the California Institute of Technology. The production of this book owes much to the typing skills of Mary Roth and, especially, to the dedicated transcription and retyping through many drafts by Shirley Arden.

PERMISSIONS

BIBLIOGRAPHY

Allison, T., and D. V. Cicchetti. "Sleep in Mammals: Ecological and Constitutional Correlates." *Science,* Vol. 149, pp. 732–734, 1976.

Arehart-Treichel, Joan. "Brain Peptides and Psychopharmacology." *Science News,* Vol. 110, pp. 202–206, 1976.

Aronson, L. R., Tobach, E., Lehrman, D. S., and J. S. Rosenblatt, eds. *Development and Evolution of Behavior: Essays in Memory of T. C. Schneirla.* W. H. Freeman, San Francisco, 1970.

Bakker, Robert T. "Dinosaur Renaissance." *Scientific American,* Vol. 232, pp. 58–72 *et seq.,* April 1975.

Bitterman, M. E. "Phyletic Differences in Learning." *American Psychologist,* Vol. 20, pp. 396–410, 1965.

Bloom, F., D. Segal, N. Ling and R. Guillemin. "Endorphins: Profound Behavioral Effects in Rats Suggest New Etiological Factors in Mental Illness." *Science,* Vol. 194, pp. 630–632, 1976.

Bogen, J. E. "The Other Side of the Brain. II. An Appositional Mind." *Bulletin Los Angeles Neurological Societies,* Vol. 34, pp. 135–162, 1969.

Bramlette, M. N. "Massive Extinctions in Biota at the End of Mesozoic Time." *Science,* Vol. 148, pp. 1696–1699, 1965.

Brand, Stewart. *Two Cybernetic Frontiers.* Random House, New York, 1974.

Brazier, M. A. B. *The Electrical Activity of the Nervous System.* Macmillan, New York, 1960.

Bronowski, Jacob. *The Ascent of Man.* Little, Brown, Boston, 1973.

Britten, R. J., and E. H. Davidson. "Gene Regulation for Higher Cells: A Theory." *Science,* Vol. 165, pp. 349–357, 1969.

Clark, W. E. Legros. *The Antecedents of Man: An Introduction to the Evolution of the Primates*. Edinburgh University Press, Edinburgh, 1959.

Colbert, Edwin. *Dinosaurs: Their Discovery and Their World*. E. P. Dutton, New York, 1961.

Cole, Sonia. *Leakey's Luck: The Life of Louis S. B. Leakey*. Harcourt Brace Jovanovich, New York, 1975.

Coppens, Yves. "The Great East African Adventure." *CNRS Research*, Vol. 3, No. 2, pp. 2–12, 1976.

Coppens, Yves, F. Clark Howell, Glynn Ll. Isaac, and Richard E. F. Leakey, eds. *Earliest Man and Environments in the Lake Rudolf Basin: Stratigraphy, Palaeoecology and Evolution*. University of Chicago Press, Chicago, 1976.

Culliton, Barbara J. "The Haemmerli Affair: Is Passive Euthanasia Murder?" *Science*, Vol. 190, pp. 1271–1275, 1975.

Cutler, Richard G. "Evolution of Human Longevity and the Genetic Complexity Governing Aging Rate." *Proceedings of the National Academy of Sciences*, Vol. 72, pp. 4664–4668, 1975.

Dement, William C. *Some Must Watch While Some Must Sleep*. W. H. Freeman, San Francisco, 1974.

DeRenzi, E., Faglioni, P., and H. Spinnler. "The Performance of Patients with Unilateral Brain Damage on Face Recognition Tasks." *Cortex*, Vol. 4, pp. 17–34, 1968.

Dewson, J. H. "Preliminary Evidence of Hemispheric Asymmetry of Auditory Function in Monkeys." In *Lateralization in the Nervous System*, S. Harnad, ed. Academic Press, New York, 1976.

Dimond, Stuart, Linda Farrington and Peter Johnson. "Differing Emotional Responses from Right and Left Hemispheres." *Nature*, Vol. 261, pp. 690–692, 1976.

Dimond, S. J., and J. G. Beaumont, eds. *Hemisphere Function in the Human Brain*. Wiley, New York, 1974.

Dobzhansky, Theodosius. *Mankind Evolving: The Evolution of the Human Species*. Yale University Press, New Haven, Conn., 1962.

Doty, Robert W. "The Brain." *Britannica Yearbook of Science and the Future*, Encyclopaedia Britannica, Chicago, 1970, pp. 34–53.

Eccles, John C. *The Understanding of the Brain*. McGraw-Hill, New York, 1973.

Eccles, John C., ed., *Brain and Conscious Experience*. Springer-Verlag, New York, 1966.

Eimerl, Sarel, and Irven DeVore. *The Primates*. Life Nature Library, Time, Inc., New York, 1965.

Farb, Peter. *Man's Rise to Civilization as Shown by the Indians of North America from Primeval Times to the Coming of the Industrial State*. E. P. Dutton, New York, 1968.

Fink, Donald G. *Computers and the Human Mind: An Introduction to Artificial Intelligence*. Doubleday Anchor Books, New York, 1966.

Frisch, John H. "Research on Primate Behavior in Japan." *American Anthropologist*, Vol. 61, pp. 584–596, 1959.

Fromm, Erich. *The Forgotten Language: An Introduction to the Understanding of Dreams, Fairy Tales and Myths*. Grove Press, New York, 1951.

Galin, D., and R. Ornstein. "Lateral Specialization of Cognitive Mode: An EEG Study." *Psychophysiology*, Vol. 9, pp. 412–418, 1972.

Gantt, Elizabeth. "Phycobilisomes: Light-Harvesting Pigment Complexes." *Bioscience*, Vol. 25, pp. 781–788, 1975.

Gardner, R. A., and Beatrice T. Gardner. "Teaching Sign-Language to a Chimpanzee." *Science*, Vol. 165, pp. 664–672, 1969.

Gazzaniga, M. S. "The Split Brain in Man." *Scientific American*, Vol. 217, pp. 24–29, 1967.

Gazzaniga, M. S. "Consistency and Diversity in Brain Organization." *Proceedings Conference on Evolution and Lateralization of the Brain, Annals of the New York Academy of Sciences*, 1977.

Gerard, Ralph W. "What Is Memory?" *Scientific American*, Vol. 189, pp. 118–126, September 1953.

Goodall, Jane. "Tool-Using and Aimed Throwing in a Community of Free-Living Chimpanzees." *Nature*, Vol. 201, pp. 1264–1266, 1964.

Gould, Stephen Jay. "This View of Life: Darwin's Untimely Burial." *Natural History*, Vol. 85, pp. 24–30, October 1976.

Gray, George W. "The Great Ravelled Knot." *Scientific American*, Vol. 179, pp. 26–39, October 1948.

Griffith, Richard M., Miyagi, Otoya, and Tago, Akira. "The Universality of Typical Dreams: Japanese vs. Americans." *American Anthropologist*, Vol. 60, pp. 1173–1179, 1958.

Grinspoon, Lester, Ewalt, J. R., and R. I. Schader. *Schizophrenia: Pharmacotherapy and Psychotherapy*. Williams & Wilkins: Baltimore, 1972.

Hamilton, C. R. "An Assessment of Hemispheric Specialization in Monkeys." *Proceedings Conference on Evolution and Lateralization of the Brain, Annals of the New York Academy of Sciences*, 1977.

Harner, M. J., ed. *Hallucinogens and Shamanism*. Oxford University Press, London, 1973.

Harris, Marvin. *Cows, Pigs, Wars and Witches: The Riddles of Culture*. Random House, New York, 1974.

Hartmann, Ernest L. *The Functions of Sleep*. Yale University Press, New Haven, Conn., 1973.

Hayes, C. *The Ape in Our House*. Harper, New York, 1951.

Herrick, C. Judson. "A Sketch of the Origin of the Cerebral Hemispheres." *Journal of Comparative Neurology*, Vol. 32, pp. 429–454, 1921.

Holloway, Ralph L. "Cranial Capacity and the Evolution of the Human Brain." *American Anthropologist*, Vol. 68, pp. 103–121, 1966.

Holloway, Ralph L. "The Evolution of the Primate Brain: Some Aspects of Quantitative Relations." *Brain Research*, Vol. 7, pp. 121–172, 1968.

Howell, F. Clark. *Early Man*. Life Nature Library, Time, Inc., New York, 1965.

Howells, William. *Mankind in the Making: The Story of Human Evolution.* Rev. ed. Doubleday, New York, 1967.

Hubel, D. H., and Wiesel, T. N. "Receptive Fields of Single Neurons in the Cat's Striate Cortex." *Journal of Physiology,* Vol. 150, pp. 91–104, 1960.

Ingram, D. "Cerebral Speech Lateralization in Young Children." *Neuropsychologia,* Vol. 13, pp. 103–105, 1975.

Jerison, H. J. *Evolution of the Brain and Intelligence.* Academic Press, New York, 1973.

Jerison, H. J. "The Theory of Encephalization." *Proceedings Conference on Evolution and Lateralization of the Brain, Annals of the New York Academy of Sciences,* 1977.

Keller, Helen. *The Story of My Life.* New York, 1902.

Korsakov, S. "On the Psychology of Microcephalics [1893]." Reprinted in the *American Journal of Mental Deficiency Research,* Vol. 4, pp. 42–47, 1957.

Kroeber, T. *Ishi in Two Worlds.* University of California Press, Berkeley, 1961.

Kurtén, Björn. *Not from the Apes: The History of Man's Origins and Evolution.* Vintage Books, New York, 1972.

La Barre, Weston. *The Human Animal.* University of Chicago Press, Chicago, 1954.

Langer, Susanne. *Philosophy in a New Key: A Study in the Symbolism of Reason, Rite and Art.* Harvard University Press, Cambridge, Mass., 1942.

Lashley, K. S. "Persistent Problems in the Evolution of Mind." *Quarterly Review of Biology,* Vol. 24, pp. 28–42, 1949.

Lashley, K. S. "In Search of the Engram." *Symposia of the Society of Experimental Biology,* Vol. 4, pp. 454–482, 1950.

Leakey, Richard E. "Hominids in Africa." *American Scientist,* Vol. 64, No. 2, p. 174, 1976.

Leakey, R. E. F., and A. C. Walker. "*Australopithecus, Homo erectus* and the Single Species Hypothesis." *Nature,* Vol. 261, pp. 572–574, 1976.

Lee, Richard, and Irven DeVore, eds. *Man, the Hunter*. Aldine, Chicago, 1968.

Le May, M., and Geschwind, N. "Hemispheric Differences in the Brains of Great Apes." *Brain Behavior and Evolution*. Vol. 11, pp. 48–52, 1975.

Lettvin, J. Y., Matturana, H. R., McCulloch, W. S., and Pitts, W. J. "What the Frog's Eye Tells the Frog's Brain." *Proceedings of the Institute of Radio Engineers*, Vol. 47, pp. 1940–1951, 1959.

Lieberman, P., Klatt, D., and W. H. Wilson. "Vocal Tract Limitations on the Vowel Repertoires of Rhesus Monkeys and Other Non-Human Primates." *Science*, Vol. 164, pp. 1185–1187, 1969.

Linden, Eugene. *Apes, Men and Language*. E. P. Dutton, New York, 1974.

Longuet-Higgins, H. C. "Perception of Melodies." *Nature*, Vol. 263, pp. 646–653, 1976.

MacLean, Paul D. *On the Evolution of Three Mentalities*, Man-Environment Systems, 1975.

MacLean, Paul, D. *A Triune Concept of the Brain and Behaviour*. University of Toronto Press, Toronto, 1973.

McCulloch, W. S., and Pitts, W. "A Logical Calculus of the Ideas Immanent in Nervous Activity." *Bulletin of Mathematical Biophysics*, Vol. 5, pp. 115–133, 1943.

McHenry, Henry. "Fossils and the Mosaic Nature of Human Evolution." *Science*, Vol. 190, pp. 425–431, 1975.

Meddis, Ray. "On the Function of Sleep." *Animal Behaviour*, Vol. 23, pp. 676–691, 1975.

Mettler, F. A. *Culture and the Structural Evolution of the Neural System*. American Museum of Natural History, New York, 1956.

Milner, Brenda, Corkin, Suzanne and Teuber, Hans-Lukas. "Further Analysis of the Hippocampal Amnesic Syndrome: 14-Year Follow-up Study of H.M." *Neuropsychologia*, Vol. 6, pp. 215–234, 1968.

Minsky, Marvin. "Artificial Intelligence." *Scientific American*, Vol. 214, pp. 19–27, 1966.

Mittwoch, Ursula. "Human Anatomy." *Nature*, Vol. 261, p. 364, 1976.

Nebes, D., and R. W. Sperry. "Hemispheric Deconnection Syndrome with Cerebral Birth Injury in the Dominant Arm Area." *Neuropsychologia*, Vol. 9, pp. 247–259, 1971.

Oxnard, C. E. "The Place of the Australopithecines in Human Evolution: Grounds for Doubt?" *Nature*, Vol. 258, pp. 389–395, 1975.

Penfield, W., and T. C. Erickson. *Epilepsy and Cerebral Localization.* Charles C. Thomas, Springfield, Ill., 1941.

Penfield, W., and L. Roberts. *Speech and Brain Mechanisms.* Princeton University Press, Princeton, N.J., 1959.

Pilbeam, David. *The Ascent of Man: An Introduction to Human Evolution.* Macmillan, New York, 1972.

Pilbeam, D., and S. J. Gould. "Size and Scaling in Human Evolution." *Science*, Vol. 186, pp. 892–901, 1974.

Platt, John R. *The Step to Man*, John Wiley, New York, 1966.

Ploog, D. W., Blitz, J., and Ploog, F. "Studies on Social and Sexual Behavior of the Squirrel Monkey (Saimari sciureus)." *Folia Primatologica*, Vol. 1, pp. 29–66, 1963.

Poliakov, G. I. *Neuron Structure of the Brain.* Harvard University Press, Cambridge, Mass., 1972.

Premack, David. "Language and Intelligence in Ape and Man," *American Scientist*, Vol. 64, pp. 674–683, 1976.

Pribram, K. H. *Languages of the Brain.* Prentice-Hall, Englewood Cliffs, N.J., 1971.

Radinsky, Leonard. "Primate Brain Evolution." *American Scientist*, Vol. 63, pp. 656–663, 1975.

Radinsky, Leonard. "Oldest Horse Brains: More Advanced than Previously Realized." *Science*, Vol. 194, pp. 626–627, 1976.

Rall, W. "Theoretical Significance of Dendritic Trees for Neuronal Input-Output Relations." In *Neural Theory and Modeling*, R. F. Reiss, ed., Stanford University Press, Stanford, 1964.

Rose, Steven. *The Conscious Brain.* Alfred A. Knopf, New York, 1973.

Rosenzweig, Mark R., Edward L. Bennett and Marian Cleeves Diamond. "Brain Changes in Response to Experience." *Scientific American*, Vol. 226, No. 2, pp. 22–29, February 1972.

Rumbaugh, D. M., Gill, T. V., and E. C. Von Glaserfeld. "Reading and Sentence Completion by a Chimpanzee." *Science*, Vol. 182, pp. 731–735, 1973.

Russell, Dale A. "A New Specimen of *Stenonychosaurus* from the Oldman Formation (Cretaceous) of Alberta." *Canadian Journal of Earth Sciences*, Vol. 6, pp. 595–612, 1969.

Russell, Dale A. "Reptilian Diversity and the Cretaceous-Tertiary Transition in North America." Geological Society of Canada Special Paper No. 13, pp. 119–136, 1973.

Sagan, Carl. *The Cosmic Connection: An Extraterrestrial Perspective.* Doubleday, New York, 1973; and Dell, New York, 1975.

Sagan, Carl, ed. *Communication with Extraterrestrial Intelligence.* MIT Press, Cambridge, Mass., 1973.

Schmitt, Francis O., Parvati Dev, and Barry H. Smith. "Electrotonic Processing of Information by Brain Cells." *Science*, Vol. 193, pp. 114–120, 1976.

Schaller, George. *The Mountain Gorilla: Ecology and Behavior.* University of Chicago Press, Chicago, 1963.

Schank, R. C., and K. M. Colby, eds. *Computer Models of Thought and Language.* W. H. Freeman, San Francisco, 1973.

Shklovskii, I. S., and Carl Sagan. *Intelligent Life in the Universe.* Dell, New York, 1967.

Snyder, F. "Toward an Evolutionary Theory of Dreaming." *American Journal of Psychiatry*, Vol. 123, pp. 121–142, 1966.

Sperry, R. W. "Perception in the Absence of the Neocortical Commissures." In *Perception and Its Disorders*, Research Publication of the Association for Research in Nervous and Mental Diseases, Vol. 48, 1970.

Stahl, Barbara J. "Early and Recent Primitive Brain Forms." *Proceedings of the Conference on Evolution and Lateralization of the Brain, Annals of the New York Academy of Sciences*, 1977.

Swanson, Carl P. *The Natural History of Man.* Prentice-Hall, Englewood Cliffs, N.J., 1973.

Teng, Evelyn Lee, Lee, P. H., Yang, K.-S., and P. C. Chang. "Handedness in a Chinese Population: Biological, Social and Pathological Factors." *Science*, Vol. 193, pp. 1148–1150, 1976.

Teuber, Hans-Lukas. "Effects of Focal Brain Injury on Human Behavior." In *The Nervous System*, Donald B. Tower, editor-in-chief, *Vol. 2: The Clinical Neurosciences.* Raven Press, New York, 1975.

Teuber, Hans-Lukas, Milner, Brenda, and Vaughan, H. G., Jr. "Persistent Anterograde Amnesia after Stab Wound of the Basal Brain." *Neuropsychologia*, Vol. 6, pp. 267–282, 1968.

Tower, D. B. "Structural and Functional Organization of Mammalian Cerebral Cortex: The Correlation of Neurone Density with Brain Size." *Journal of Comparative Neurology*, Vol. 101, pp. 19–51, 1954.

Trotter, Robert J. "Language Evolving, Part II." *Science News*, Vol. 108, pp. 378–383, 1975.

Trotter, Robert J. "Sinister Psychology." *Science News*, Vol. 106, pp. 220–222, October 5, 1974.

Turkewitz, Gerald. "The Development of Lateral Differentiation in the Human Infant." *Proceedings of the Conference on Evolution and Lateralization of the Brain, Annals of the New York Academy of Sciences*, 1977.

Vacroux, A. "Microcomputers." *Scientific American*, Vol. 232, pp. 32–40, May 1975.

Van Lawick-Goodall, Jane. *In the Shadow of Man.* Houghton-Mifflin, Boston, 1971.

van Valen, Leigh. "Brain Size and Intelligence in Man." *American Journal of Physical Anthropology*, Vol. 40, pp. 417–424, 1974.

Von Neumann, John. *The Computer and the Brain.* Yale University Press, New Haven, Conn., 1958.

Wallace, Patricia. "Unravelling the Mechanism of Memory." *Science*, Vol. 190, pp. 1076–1078, 1975.

Warren, J. M. "Possibly Unique Characteristics of Learning by Primates." *Journal of Human Evolution,* Vol. 3, pp. 445–454, 1974.

Washburn, Sherwood L. "Tools and Human Evolution." *Scientific American,* Vol. 203, pp. 62–75, September 1960.

Washburn, S. L., and R. Moore. *Ape Into Man.* Little, Brown, Boston, 1974.

Webb, W. B. *Sleep, The Gentle Tyrant.* Prentice-Hall, Englewood Cliffs, N.J., 1975.

Weizenbaum, Joseph. "Conversations with a Mechanical Psychiatrist." *The Harvard Review,* Vol. 111, No. 2, pp. 68–73, 1965.

Wendt, Herbert. *In Search of Adam.* Collier Books, New York, 1963.

Witelson, S. F., and W. Pallie. "Left Hemisphere Specialization for Language in the Newborn: Neuroanatomical Evidence of Asymmetry." *Brain,* Vol. 96, pp. 641–646, 1973.

Yeni-Komshian, G. H., and D. A. Benson. "Anatomical Study of Cerebral Asymmetry in the Temporal Lobe of Humans, Chimpanzees, and Rhesus Monkeys." *Science,* Vol. 192, pp. 387–389, 1976.

Young, J. Z. *A Model of the Brain.* Clarendon Press, Oxford, 1964.

GLOSSARY

Accessing Computer jargon for making contact with information stored elsewhere.

Affect (noun) A feeling or emotion, particularly a strong one.

Alexia A weakening or loss of the ability to comprehend written or printed words and sentences. Compare with aphasia.

Ameslan American sign language, widely used by persons with impaired speech and hearing.

Amygdala An almond-shaped component of the limbic system adjoining the temporal lobe of the neocortex.

Anaglyph A two-dimensional stereo representation of a three-dimensional image; most often composed of red and green dots, and viewed with red and green eyeglasses.

Anterior commissure A relatively minor bundle of nerve fibers that connects the left and right cerebral hemispheres of the neocortex. Compare with corpus callosum.

Aphasia Generally, a weakening or loss of the ability to articulate ideas by language in any form. It is sometimes used more narrowly to indicate the inability to recognize spoken words. Compare with alexia.

Bilateral On both sides.

Bits Units of binary information. One bit is the answer to a single yes or no question.

Brainstem See Hindbrain.

Broca's area A portion of the neocortex intimately connected with speech.

Buffer dumping The accessing (q.v.) or disposal of information temporarily deposited in a short-term memory.

CC Abbreviation for cubic centimeter.

Cerebellum A brain mass lying in the back of the head underneath the posterior cerebral cortex and above the pons and medulla in the hindbrain. Like the neocortex, it has two hemispheres.

Cerebral cortex In humans and higher mammals the large outer layer of the cerebral hemispheres, in major part responsible for our characteristically human behavior. Sometimes synonymous with neopallium or neocortex (q.v.).

Cetacea An order of aquatic mammals that includes whales and dolphins.

Chromosomes The long strands of hereditary material containing the genes, and composed exclusively of nucleic acids.

Convolution See Gyrus.

Corpus callosum The great commissure, or bundle of nerve fibers, which is the principal cabling between the left and right hemispheres of the cerebral cortex.

Craniotomy The cutting or removal of part of the skull, generally as an antecedent to brain surgery.

DNA Deoxyribonucleic acid. See Nucleic acids.

Electrode A solid electrical conductor through which an electric current moves. Electrical currents in the brain are sensed by an electroencephalograph through its electrodes.

Electroencephalograph (EEG) A device consisting of amplifiers and a pen automatically writing on a rotating drum, used for recording the electrical currents in the brain conducted to the device by electrodes attached to the surface of the head. It is useful for medical diagnosis and for studies of the function of the brain.

Endocast A mold of an interior; in this book, a mold of the interior of a fossil braincase.

Endocranial Within the skull.

Endorphins Small internally produced brain proteins which can induce a variety of emotional or other states in animals.

Equipotent Having equal ability; in particular, the view that for certain cognitive or other functions any part of the brain can substitute for any other.

Extirpation The entire removal of a unit of the brain, usually by surgical procedures.

Extragenetic information Information carried outside the genes—generally in brains and in cultures.

Extrasomatic information Information carried outside the body (for example, the contents of books).

Forebrain The evolutionarily most recent of the three major divisions of the vertebrate brain. Also called the prosencephalon. It is divided in turn into the R-complex, limbic system, and neocortex.

Frontal lobe Approximately, the portion of the neocortex beneath the forehead.

Gametes Mature sperm or egg cells capable of participating in fertilization. They contain a haploid (q.v.) number of chromosomes.

Gyrus One of the prominent rounded elevations on the surface of the neocortex. Also called convolution.

Haploid Having a number of chromosomes equal to half the number in an ordinary body or somatic cell. For example, in human beings each somatic cell has 46 chromosomes but each gamete (q.v.) has 23 chromosomes.

Hindbrain The most ancient part of the brain, including the pons, cerebellum, medulla oblongata, and the upper portion of the spinal cord. It is also called the brainstem or the rhombencephalon.

Hippocampal commissure A relatively minor bundle of nerve fibers which connects the left and right hemispheres of the cerebral cortex near the hippocampus. Compare with corpus callosum.

Hippocampus A structure in the limbic system connected with memory.

Hypothalamus A portion of the limbic system lying below the thalamus which, among other functions, helps to regulate bodily temperature and metabolic processes.

KG Abbreviation for kilogram.

Lateralization The separation of function between two sides, especially the left and right hemispheres of the neocortex.

Lesion A cut, wound, or injury. Some brain lesions occur by accident and some by surgical procedure.

Limbic system The part of the forebrain intermediate in locale and antiquity between the R-complex and the neocortex.

Lobes of the neocortex See Frontal lobe, Occipital lobe, Parietal lobe, and Temporal lobe.

Lobotomy A surgical incision into or lesion of one of the neocortical lobes (q.v.).

Localization of brain function The finding that certain parts of the brain perform certain specific functions. It is the opposite of the equipotent (q.v.) hypothesis.

Long-term memory Memory retained for substantial periods of time—for example, more than a day.

M Abbreviation for meter.

Medulla oblongata (sometimes called simply **Medulla**) The portion of the brain at the region of its connection with the spinal cord. It is a part of the hindbrain.

Microcephalic One with an abnormally small head. The condition is often associated with significant mental impairment.

Midbrain The middle region of the vertebrate brain, between the hindbrain and forebrain. Also called the mesencephalon.

Motor cortex The portion of the neocortex concerned with motion and coordination of the limbs.

Mutations Inheritable changes in the nucleic acids of chromosomes.

Natural selection The principal method of biological evolution, as first described by Darwin and Wallace. The preferential survival and reproduction of organisms fortuitously better adapted to their environments than their competitors.

Neocortex The outermost, evolutionarily most recent part of the cerebral cortex. Sometimes used as synonymous with cerebral cortex.

Neural chassis The combination of spinal cord, hindbrain, and midbrain.

Neuron or **Neurone** A nerve cell, the basic unit of the nervous system, and the fundamental building block of the brain.

Niche, ecological An organism's role in nature.

Nucleic acids The genetic material of all life on Earth, consisting of ladder-like sequences of units called nucleotides, usually arranged in a double helix. There are two main varieties of nucleic acids, DNA and RNA.

Nucleotide The fundamental building blocks of the nucleic acids (q.v.).

Occipital lobe Approximately, the portion of the neocortex under the back of the skull.

Olfactory bulbs Components of the brain attached to the front of the forebrain, and playing an important role in the perception of smells.

Parietal lobe Aproximately, the middle portion of each cerebral hemisphere of the neocortex.

Pituitary The "master" endocrine gland, situated in the limbic system but near the midbrain and influencing both growth and the operations of other endocrine glands.

Plasticity The capability to be shaped or formed; in particular, the ability to learn from the external environment.

Pons (also called **Pons varioli**) The neural bridge connecting the medulla oblongata and the midbrain. It is a part of the brainstem.

Prewired Computer jargon for information already in place. Also called hard-wired. The more prewiring, the less plasticity.

Primary processes Psychoanalytic term for the fundamental unconscious functions of the brain.

Primates An order (one of the taxonomic classifications) of mammals that includes lemurs, monkeys, apes, and humans.

Proteins Along with the nucleic acids, the principal molecular basis of life on Earth. Proteins are made of constituent units called amino acids and are ordinarily elaborately folded and coiled. Some proteins are spherical in overall shape, while others resemble free-standing nonrepresentationalist sculpture. All enzymes, which control the rate of chemical reactions in the cell, are proteins. The synthesis and activation of enzymes are controlled by the nucleic acids.

Psychomotor Relating to mental control of muscular processes.

R-complex or **Reptilian complex** The evolutionarily most ancient part of the forebrain.

Recapitulation or **the recapitulation of phylogeny by ontogeny**
The apparent repetition, during the embryonic development of an
individual organism, of a past evolutionary stage of the species.

REM Rapid eye movements, particularly those which occur under
the eyelids during dream sleep. Therefore, the characterization of
such a sleep.

RNA Ribonucleic acid. See Nucleic acids.

Selection pressure In evolutionary theory, the influence of the environment in selecting for survival and reproduction a particular set
of genetic characteristics.

Short-term memory Memory retained for brief periods of time—
for example, less than a day.

Synapse The junction of two neurons: the locale where an electrical impulse is transmitted from one neuron to another.

Taxon (plural, **Taxa**) A group of organisms classified according to
common characteristics, ranging from minor distinctions such as
races and subspecies to major distinctions such as the differences
between the plant and animal kingdoms.

Temporal lobe Approximately, the portion of the neocortex beneath
the temples of the skull.

Thalamus A portion of the limbic system near the center of the
brain. Among other functions, it replays sensory stimuli to the neocortex.

– – – – **tomy** The cutting of an organ represented by the dashes
(see, for example, Craniotomy or Lobotomy).

Triune brain The idea, most recently advocated by Paul MacLean, that the forebrain comprises three separately evolved and to some degree independently functioning cognitive systems.

Zygote A fertilized egg.

INDEX